Applications of NDT Data Fusion

Applications of NDT Data Fusion

edited by

Xavier E. Gros

DUT, BSc (Hon), MSc, PhD

Independent NDT Centre
Bruges
France

Springer Science+Business Media, LLC

Library of Congress Cataloging-in-Publication Data

Applications of NDT data fusion / edited by Xavier E. Gros.
p. cm.
ISBN 978-1-4613-5540-3 ISBN 978-1-4615-1411-4 (eBook)
DOI 10.1007/978-1-4615-1411-4
1. Non-destructive testing--Data processing. 2. Multisensor data fusion. I. Gros, X.E. (Xavier Emmanuel), 1968-

TA417.2.A68 2001
620.1'127--dc21

2001029934

Originally published by Kluwer Academic Publishers in **2001**
Softcover reprint of the hardcover 1st edition **2001**

Printed on acid-free paper.

Contents

LIST OF CONTRIBUTORS

Daniel Babot
CNDRI, INSA 303, 69621 Villeurbanne Cedex, France
cndribab@insa-lyon.fr

Richard H. Bossi
Boeing Defense & Space Group, PO Box 3999 MS 8H-05, Seattle, WA 98124-2499, USA
richard.h.bossi@boeing.com

Uwe Fiedler
Fraunhofer IzfP, Einrichtung EADQ, Kruegerstrasse 22, 01326 Dresden, Germany
fiedler@eadq.izfp.fhg.de

David S. Forsyth
Institute for Aerospace Research, National Research Council Canada, Building M14, Montreal Road, Ottawa, ON, Canada K1A 0R6
david.forsyth@nrc.ca

Pan Fu
Department of Aerospace, Civil and Mechanical Engineering, University of Hertfordshire, Hatfield Campus, College Lane, Hatfield AL10 9AB, England, UK
pan_fu@hotmail.com

Bernard Gibaud
Laboratoire IDM/SIM, Université de Rennes 1, Faculté de Medecine, 2 avenue Professeur Léon Bernard, 35043 Rennes Cedex, France
bernard.gibaud@univ-rennes1.fr

Xavier E. Gros
Independent NDT Centre, 14 rue G.Brassens, 33520 Bruges, France
xgros@ieee.org

Christophe Grova
Laboratoire IDM/SIM, Université de Rennes 1, Faculté de Medecine, 2 avenue Professeur Léon Bernard, 35043 Rennes Cedex, France
christophe.grova@univ-rennes1.fr

Anthony D. Hope
Design and Advanced Technology Research Centre, Faculty of Technology, Southampton Institute, East Park Terrace, Southampton SO14 0RD, England, UK
tony_hope@solent.ac.uk

Seth Hutchinson
Dept. of Electrical and Computer Engineering, University of Illinois, The Beckman Institute, 405 North Mathews Ave., Urbana, IL 61801, USA
seth@uiuc.edu

Pierre Jannin
Laboratoire IDM/SIM, Université de Rennes 1, Faculté de Medecine, 2 avenue Professeur Léon Bernard, 35043 Rennes Cedex, France
pierre.jannin@univ-rennes1.fr

Valérie Kaftandjian
CNDRI, INSA 303, 69621 Villeurbanne Cedex, France
valkaf@insa.insa-lyon.fr

Jerzy P. Komorowski
Institute for Aerospace Research, National Research Council Canada, Building M14, Montreal Road, Ottawa, ON, Canada K1A 0R6
jerzy.komorowski@nrc.ca

James M. Nelson
Boeing Mathematics and Computing Technology, PO Box 3999 MS 7L-25, Seattle, WA 98124-2499, USA
james.m.nelson@boeing.com

James Reed
University of Michigan, School of Information, 304 West Hall, Ann Arbor MI 48109-1092, USA
jreed@umich.edu

Liu Zheng
School of Electrical and Electronic Engineering, Block S1-B4C-14, Intelligent Machines Research Laboratory, Control & Instrumentation Division, Nanyang Technological University, Nanyang Avenue, Singapore 639798
liu@linuxstart.com

Yue Min Zhu
CREATIS-CNRS, Research Unit 5515 affiliated to INSERM, INSA 502, 69621 Villeurbanne Cedex, France
zhu@creatis.insa-lyon.fr

ACKNOWLEDGEMENTS

The editor would like to thank all contributing authors for accepting to write about their work, and without whom this book could not have been completed.

Special thanks go to David Gilbert of the British Institute of NDT (Editor of *Insight*) for accepting to publish early articles on NDT data fusion at a time when scepticism on this topic was high among members of the NDT community. This book is evidence that *NDT Data Fusion* is now a valuable and applied process.

Thanks are also due to C. Pecorari (ABB Corporate Research, Sweden) and R.D. Wakefield (The Robert Gordon University, Scotland) for their critics and constructive comments. The help in programming of Julien Bousigué (Ingénieur E.N.S.I. de Caen, D.E.A. Traitement et Synthèse d'Image) is also acknowledged.

LIST OF ABBREVIATIONS

The acronyms and abbreviations listed below are not definitive and are given only for information as they are used in this book.

AC	Alternating Current
AE	Acoustic Emission
AFRL	Airforce Research Laboratory
AI	Artificial Intelligence
ALC	Air Logistics Center
A-Scan	Amplitude Scan
ASCII	American Standard Code for Information Interchange
AVS	Application Visual Software
Bel	Belief
bpa	Basic Probability Assignment
B-Scan	Brightness Scan
CAD	Computer Aided Design
CAI	Composites Affordability Initiative
C-C	Carbon-Carbon
CCD	Charge Coupled Device
CFRP	Carbon Fibre Reinforced Plastic
C-Scan	Contrast Scan
CT	Computer Tomography
dB	Decibel
DB4	Daubechi Wavelet 4
DS	Dempster-Shafer
DSA	Digital Subtraction Angiography
EC	Eddy Current
ECG	Electro-Cardiography
ECoG	Electro-Corticography
EEG	Electro-Encephalography
EMG	Electro-Myography
EOL	Edge Of Light
EV	Enhanced Visual
FCM	Fuzzy C-Means
FE	Finite Element
FEM	Finite Element Modelling
FFT	Fast Fourier Transform
FLIR	Forward Looking Infrared
fMRI	Functional Magnetic Resonance Imaging
GHA	Generalised Hebbian Algorithm
GPR	Ground Penetrating Radar
HOS	Higher-Order Statistics
Hz	Hertz
IFFT	Inverse Fast Fourier Transform
INDERS	Integrated NDE Data Evaluation and Reduction System
IRT	Infrared Thermography
IUS	Inertial Upper Stage
JANNAF	Joint Army Navy NASA Airforce
LOI	Lift-Off Point of Intersection

LUT	Laser Ultrasonics
MEG	Magneto-Encephalography
MLP	Multilayer Perceptron
MRA	Magnetic Resonance Angiography
MRI	Magnetic Resonance Imaging
MSFC	Marshall Space Flight Center
NDE	Non-Destructive Evaluation
NDI	Non-Destructive Inspection
NDT	Non-Destructive Testing
NE	Network Editor
NMR	Nuclear Magnetic Resonance
NN	Neural Network
NURBS	Non-Uniform Rational B-Splines
O&M	Operations & Maintenance
OOP	Object Oriented Programming
PC	Personal Computer
PCA	Principal Component Analysis
PET	Positron Emission Tomography
PFA	Probability of False Alarm
PFC	Probability of False Calls
Pls	Plausibility
POD	Probability of Detection
RBF	Radial Basis Function
ROC	Receiver (or Reliability) Operating Characteristic
ROV	Remotely Operated Vehicle
SAFT	Synthetic Aperture Focusing Technique
SEEG	Stereo-Electro-Encephalography
S/N	Signal-to-Noise
SNR	Signal-to-Noise Ratio
SPECT	Single Photon Emission Computed Tomography
SPIP	Solid Propulsion Integrity Program
3-D	Three Dimension(al)
TOFD	Time-Of-Flight Diffraction
2-D	Two Dimension(al)
UCD	Unstructured Cell Data
US	Ultrasound
UT	Ultrasonic Testing
VRML	Virtual Reality Modelling Language
WVD	Wigner-Ville Distribution
WT	Wavelet Transform
XRII	X-Ray Image Intensifier

NOTE

Colour figures – printed in black & white in this book – can be viewed on the World Wide Web at:

http://www.geocities.com/xgros/

following the link to *Applications of NDT Data Fusion.*

1 MULTISENSOR DATA FUSION AND INTEGRATION IN NDT

Xavier E. GROS
Independent NDT Centre,
Bruges, France

1.1 INTRODUCTION

In a previous publication the concept of data fusion applied to non-destructive testing (NDT) was introduced [1]. The present book explores the concept of NDT data fusion through a comprehensive review and analysis of current applications. Its main objective is to provide the NDT community with an up-to-date publication containing writings by authoritative researchers in the field of data fusion. It is not intended to give rigorous scientific details, but more a pragmatic overview of several applications of data fusion for materials evaluation and condition monitoring.

This publication arrives at a time when scepticism and lack of knowledge by decision-making people in the NDT community are reluctant to adopt new ideas. Indeed, it is sometimes believed that the concept of data fusion is a buzzword with no future and substantial applications. The present book not only indicates that data fusion is applied in NDT, but also that it is becoming a major tool in industrial research and development. In 1999, a section exclusively dedicated to data fusion was organised at the famous 'Progress in Quantitative Nondestructive Evaluation' conference. The new journal 'Information Fusion', a scientific magazine aimed at describing theoretical and experimental applications of data fusion, is another evidence of increasing activities in this field.

No single NDT technique or method can fully assess the structural integrity of a material. Indeed, each method presents some limitations in terms of defect detection and characterisation. Inspection apparatus can generate incomplete, incorrect or conflicting information about a flaw or a defect. Additionally, poor signal-to-noise ratio may make signal interpretation complex or unreliable. Therefore, the use of more than one NDT system is usually required for accurate defect detection and/or quantification. In addition to a reduction in inspection time, important cost saving could be achieved if a data fusion process is developed to combine signals from multisensor systems for manual and remotely operated inspections. For these reasons, the concept of NDT data fusion - based on the synergistic use of information from multiple sources in order to facilitate signal interpretation and increase the probability of defect detection and accuracy in characterisation - is expanding rapidly.

There is no denying that the data fusion concept is relatively simple but its implementation remains challenging. Multisensor fusion refers to the acquisition, processing and synergistic combination of information gathered by various knowledge sensors to provide a better understanding of the phenomenon under consideration [2-3]. Despite relatively fast progress in NDT data fusion, several problems remain to be solved. There is no magic formula or fusion algorithm that will allow one to carry out fusion of multiple data sets. It is an on-going process, a technology that is evolving and adapting to the requirements and needs of the moment. Researchers have to deal with incomplete data, different data formats, and registration problem [4]. Both the fusion architecture and the mathematical tools to select are also dependent on the fusion goals (e.g. defect detection, flaw sizing).

Although the concept of multisensor data fusion and integration has been applied to military applications for several years, only recently did the NDT community realise that it could also be of benefit to the field of non-destructive evaluation (NDE). When the terminology related to NDT data fusion was first introduced in 1994, the first reaction was to ignore this new concept as it was thought too complex [5-6]. But, with time and due to the growing interest of the aerospace and nuclear industries (both in need of accurate and efficient measurements), information fusion became more easily accepted and started to move from the laboratory to on-site testing. At the moment, and judging by the number of publications in this topic, Europe, and more particularly France, shows a particular interest in developing fusion systems. Integration of inspection results with CAD images of the structure inspected is a recent development that facilitates defect location and simplifies diagnosis [7]. A recent survey indicated that industrial engineering application of information fusion is the field with fewer publications (5%) mainly because it is a field in emergence for information fusion [8].

The use of multiple sensors for NDT and the integration of information from these sensors have been applied as early as 1989 [9]. Computer tomography, eddy current and ultrasonic data were interpolated to a common pixel raster format for comparison but no data fusion operation was performed. Since then, new advances in the field of NDT signal processing and NDT data fusion have been accomplished. One of the most obvious examples is probably in the field of image processing and computer visualisation. What was once considered state-of-the-art technology for the enhancement of radiographs [10] may appear old fashioned compared to nowadays real time dynamic virtual reality systems [11]. Virtual reality is used for training and simulation of remotely operated inspections in hazardous areas. Remotely operated vehicles (ROVs) are piloted in a virtual reality mode and their progress displayed on a computer screen showing both the environment in which the system performs and the inspection data [11]. Combined with data fusion technology and developments in artificial intelligence, fully autonomous ROVs for underwater inspection of pipelines could soon become reality [12].

The development of new software and techniques has lead to a rapid evolution of the applications of NDT data fusion.

1.2 OVERVIEW OF CHAPTERS

Statistic and probabilistic methods are being used increasingly in NDT to evaluate the performance of an inspection system or an operator [1]. The use of probability of detection (POD) and receiver operating characteristic (ROC) curves have been successfully applied to existing material manufacturing and evaluation processes for the improvement of quality and process control [13]. The use of ROC curves to evaluate the potential of data fusion processes is described in chapter 2.

Examples of pixel level image fusion applied to the non-destructive inspection (NDI) of composite panels used by the aerospace industry are given in chapter 3. Results from the fusion of images from multiple NDT sources gathered during the inspection of composite materials damaged by impact are presented. Pixel level data fusion evolved from simple image fusion operations (e.g. AND, OR operations) to pyramid decomposition based fusion, wavelet transform based fusion and multi-resolution mosaic technique. Image fusion has been thoroughly investigated by researchers world-wide, and led to various fusion algorithms for image fusion and pixel level image fusion. Each publication presenting a specific algorithm or application. Most algorithms developed for military, robotics, or geoscience applications can be adapted for the fusion of inspection data. For example, Lallier and Farooq described a real time pixel level image fusion process via adaptive weight averaging that could be adequate for on-line inspection [14]. Petrović and Xydeas presented a novel efficient approach to image fusion using a multiscale image processing technique [15]. Fusion is achieved by combining image information at two different ranges of scales with spectral decomposition being performed using adaptive size averaging templates. The particularity of their approach is such that each feature of an image is fused using a specific fusion algorithm depending on its size. A possible application could be the fusion of images containing defects of various sizes.

As reported by Fu and Hope in chapter 4, data fusion also finds application in condition monitoring. Because effective machine's health assessment requires processing of a large amount of data, the concept of data fusion is well suited to deal with such task. Hope and Fu applied artificial intelligence technology for condition monitoring in an industrial milling environment [16]. They showed that a fuzzy pattern recognition approach could be used to monitor tool wear. Their system performs feature extraction, establishment of membership function of all the features of standard models, establishment recognition criteria, and classification operation. Goebel et al. developed a data fusion system that performs a weighted average of individual tools using confidence values assigned dynamically to each individual diagnostic tool [17]. Application is for condition monitoring of high speed milling machines. The authors concentrated on tool wear because it is a highly non-linear process that is hard to monitor and estimate. The fusion of diagnostic estimates is performed via fuzzy validation curves, and diagnostics were established based on data from three different types of sensors (i.e. acoustic emission, vibration, motor current).

An approach to 3-D image reconstruction from a limited number of views is presented in chapter 5. Here, the authors show that it is possible to combine images from ultrasonic C and D-scans gathered using a system with limited capabilities. Detected defects were presented in a non-continuous pixel manner by the apparatus. Image processing operations (e.g. thresholding, Gaussian filtering, Canny edge detection process for contour mapping, etc.) helped produce a fused image that reveals the actual defect shape in 3-D [18]. Pastorino also discussed image reconstruction to recover the shape of a defect or of a structure from a limited number of views obtained with microwave testing, computer tomography (CT), radiography and ultrasounds [19]. Summa *et al.* developed an interactive 3-D visualisation and image analysis package for processing of NDI data [20]. Their software allows data import, rendering and manipulation of full waveform data from several kinds of commercial ultrasonic equipment. Volume rendering makes accessible view angles not normally reachable from the outside, and can be useful to identify features that may not be readily apparent on a set of sequential 2-D views because of their relative size, location or orientation. Using a 3-D computer model, scientists can do a virtual dissection of the part, stripping away exterior layers of dissimilar materials to reveal hidden structures, or to examine shape and texture of interior surfaces. Defect localisation and sizing in 3-D images is improved. Animation and slicing can also provide additional information. The potential of 3-D image reconstruction was also demonstrated to generate an image of flaws from a limited number of cone-beam X-ray projections. Vengrinovich *et al.* used the Bayesian reconstruction with Gibbs prior in the form of mechanical models like non-causal Markov fields [21].

Another application of data fusion is for the inspection of electronic components. In chapter 6 Hutchinson and Reed describe image fusion with subpixel edge detection and parameter estimation for automated optical inspection of electronic components [22]. An image sequence is captured in a digital form during the movement of an electronic component in front of a camera. A high-resolution image is generated from the fusion of this sequence of images. A set of data points is then produced by subpixel edge detection. An ellipse parameter estimation procedure back-projects this set of data points into 3-D space to estimate the size of inspected circular features. The author developed an algorithm that is well adapted to determine features of small size holes used for mounting integrated circuits and other electronic components.

Chapter 7 deals with NDT data fusion applications for the aerospace industry. Bossi and Nelson report research performed at Boeing for NDE multimode data fusion of aircraft parts. They showed that combining multiple techniques and fusing these results on a CAD image of the inspected component facilitate signal interpretation [23]. Similarly, Matuszewski *et al.* described three fusion approaches to combine ultrasonic, radiographic and shearographic images of an aircraft part [24]. The superiority of the region-based fusion approach for defect interpretation was explained by the fact that the fusion parameters were optimised according to the unique geometrical and material structure contained in the different parts of the inspected component.

Data fusion applied to weld inspection is described in chapter 8. The authors fused radiographic and ultrasonic images using Dempster-Shafer evidential theory. By exploiting complementary information from two different sources (i.e. defect morphology provided by radiography and defect depth information gained with ultrasounds) the overall knowledge of a defect and the extent of weld damage can be evaluated.

Chapter 9 contains information on data fusion in civil engineering. Information fusion is carried out for decision support and to separate artefacts from the magnetic signal gathered during bridge examination.

NDT data fusion for improved corrosion detection is described in chapter 10. Forsyth and Komorowski applied data fusion to enhance corrosion detection and signal interpretation in aircraft lap joints. Corrosion is an important issue in NDT and data fusion is well suited to overcome some of the limitations in accurately detecting and quantifying corrosion. Siegel and Gunatilake developed a sensor fusion approach to detect surface cracks, surface corrosion and subsurface corrosion based on information gathered by visual cameras mounted on a mobile robot [25]. A wavelet transform operation of a visually acquired image prior to feature extraction in scale space and defect classification was performed.

The book concludes with a chapter on applications of data fusion for medical applications. This is an area in continuous expansion, as most tests in the biomedical field are carried-out non-destructively for obvious reasons. Such tests also require as much information as possible from complementary techniques to help doctors establish an accurate diagnosis. Since biomedical imaging requires the use of multiple techniques, the medical field is one the most challenging and fast growing areas in which data fusion is developed and applied. Radiography is used to detect broken bones, CT to detect physiological anomalies, ultrasound for pregnancy tests, nuclear magnetic resonance (NMR or MRI) to measure electromagnetic field variations of diseased tissues, etc. Test interpretation is difficult because the human body is made of heterogeneous materials, and difficulties arise due to optical dispersion or diffraction in the case of optical spectroscopy for example [26]. In medical applications, several software programmes (e.g. Voxar Plug'n View 3D) make use of fusion techniques to display and render images from different modalities on a unique reference model. Medical applications that combine radiographic and CT images are increasing. Ault and Siegel developed a data fusion system to combine ultrasonic images with multiple CT cross-section images and reconstruct a 3-D picture of a bone [27]. Ultrasonic images can be digitally fused with laser Doppler blood-flow measurements to identify the precise location of internal bleeding [28]. Fusion techniques have also been developed to register medical data with respect to a surface fitting method, and for surface contour reconstruction to help in pre-surgical planning and radiation treatment [29]. Reconstruction of vertebra to monitor spinal diseases was achieved by Gautier *et al.* through low level fusion of parallel MRI slices [30]. A segmentation operation based on an active contour method is performed to establish the contour of a vertebra. To reduce conflict of information, active contours are classified using the theory of evidence, taking into consideration knowledge of the geometry of the segmented object. Next, a low level image fusion operation is performed to separate cortex from air and fluid. Finally, a fused image with belief information is obtained for improved decision. Recently a French company developed software for registration and fusion of data from medical applications that allows enhancement of signal analysis and facilitates diagnosis [31]. Such a system is in use in French medical establishments and the benefits gained from applying data fusion were of tremendous value.

1.3 CURRENT AND FUTURE TRENDS

It would be difficult to compile an exhaustive list of all current and future applications of data fusion for condition monitoring and NDE. Indeed, recent research efforts have focused on improving decision support, facilitating sensors' management, estimating probabilities of defect characteristics, fusing images for 3-D reconstruction and data fusion for autonomous remote inspection systems.

Probably one of the most recent industrial applications was the fusion of ultrasonic and radiographic data for improved automated defect detection in welds [32]. The fusion algorithm is based on the Dempster-Shafer theory, and the representation of uncertainty in a colour-coded manner facilitates signal analysis. Although defects of small dimensions could be accurately detected, it was clear that false alarms could not be avoided.

European industries are working closely with research institutes to develop data fusion systems that will suit specific applications. The oil company OIS applied data fusion for the inspection of welds by fusing ultrasonic and radiographic images [33]. The MISTRAL project funded by the European Commission involves several European industrials and is aimed to design, develop and evaluate multisensor approaches on welded components shared by several companies. It consists of multi-technique inspection probes, processing tools, a fracture mechanics code in conjunction with standard acceptance criteria, and a fusion procedure of signals from complementary NDT techniques [34]. A consortium of industrials and researchers investigated the development and use of data fusion for quality checking of wood-based panels and solid wood. Their work concentrated on a control system based on fuzzy logic for board production and the development of a prototype system for visual detection of surface defects in wood and wood based materials. The results of the project have been used subsequently by commercial enterprises. An objective of a BRITE/EURAM project (project number: BRPR960289) is to reduce inspection cost by advanced data fusion and diagnosis data integration for the aeronautical industry. Another deals with data management and fusion of signals from a multi-sensor inspection system for application in chemical and energy industries (e.g. NDT of heat exchangers, pressure vessels, and pipes).

Another application is the detection of local inhomogeneities in composites [35]. By using defect location determined with acoustic emission and segmenting ultrasonic and radiographic images, Jain *et al.* used a fusion technique to reconstruct a complete map of the defect location and shape [35]. Image segmentation is performed for defect extraction from noisy images (e.g. poor contrast radiographs). They applied the Yanowitz and Bruckstein method [36] to construct a thresholding surface by interpolating the image grey levels at points where the gradient is high, indicating probable object edges. Registration problems were solved by a neighbouring pixel approach.

The use of neural networks to perform data fusion operations is another aspect worth mentioning. Researchers in the USA used multi-layer perceptron (MLP) and radial basis function (RBF) networks to fuse ultrasonic and eddy current information at the signal, pixel, feature or symbolic level [37]. Barniv and Casasent used the correlation coefficient between pixels in the grey level images as a basis for registering images from multiple sensors [37], while Akerman used the pyramidal technique [38], and Haberstroh and Kadar a MLP for multi-spectral data fusion [39]. Applying artificial intelligence technology to the field of NDT appears very promising, and could substitute the human operator when making impartial repair decisions. Other researchers developed a scheme based on defect impulse response using robust deconvolution techniques based on higher-order statistics (HOS) [40]. HOS preserves the phase character of the flaw signature before a Wigner-Ville distribution (WVD) transforms the impulse response into a time-frequency image. The flaw signature contained in each signal is determined by performing principal component analysis (PCA) in an adaptive fashion using an unsupervised artificial neural network block trained through the generalised Hebbian algorithm (GHA). The PCA blocks perform high dimensionality reduction to extract optimal features from the WVD images. The extracted features are fed into a MLP network for defect classification among five different defect classes. Wang and Cannon presented a flexible manufacturing paradigm in which robot grasping is interactively specified and skeletal images used for rapid surface flaw identification tasks at the manufacturing level [41]. A novel material handling approach is described for robotic picking and placing parts onto an inspection table using virtual tools and a cyber-glove for virtual interaction with the part to be inspected. Image analysis is also performed for defect identification and a neural network is trained to recognise gestures that control movement of the part. The cyber-glove is used to convert operator hand motions into virtual robot gripper motions as well as camera motions [41].

Hannah *et al.* demonstrated the strategy and structures involved in making decisions based on condition data applied to the fields of combustion and fault diagnostics analysis [42]. Fuzzy membership functions are applied in data association, evaluation of alternative hypotheses in multiple hypothesis testing, fuzzy-logic based pattern recognition, and fuzzy inference schemes for sensor resource allocation. Stover *et al.* developed a general purpose fuzzy-logic architecture in robotics that controls sensing resources, fusion of data for tracking, automatic object recognition, system resources and elements and automated situation assessment [43]. Fuzzy techniques can also be used for image analysis and to reveal information of key interest to NDT operators [44].

Increasing applications of NDT data fusion led to the development of new tools necessary to combine information efficiently (e.g. software, fusion architecture, uniform data format). For example, the Unix based CIVA software was developed for processing of eddy current, radiographic and ultrasonic data, as well as data fusion and 3-D visualisation [45]. It allows a wide range of image and signal manipulations such as defect extraction, contour of defect, mapping, 3-D visualisation, images fusion, defect sizing and segmentation. CIVA software is an open system for processing NDE data, and has the ability to read different file formats [46]. Additional features include image superposition (e.g. CAD images with inspection results). It was claimed to be a fully adaptable software that incorporates different modelling programs for field applications.

The TRAPPIST system (Transfer, Processing and Interpretation of 3-D NDT Data in a Standard Environment) was designed by European research teams to improve the accuracy of NDT signal interpretation using an automated expert system [47]. It targets applications in the domain of general industry and manufacturing. They studied the combination of 3-D NDT under various aspects and implemented a system that combined data sources from remote locations. The use of this technology for different inspection tasks encourages the acceptance for combined decentralised test systems. This led to standardised data formats and co-ordinate systems to deal with signals from different physical tests. The TRAPPIST concept has created an awareness of the benefits of integrating several methods for quality assurance processes by means of telecommunications [47]. One achievement of the TRAPPIST project was to combine radiographic and ultrasonic images to reveal the exact geometry of bore holes and to distinguish between bore holes and delamination otherwise impossible to recognise from individual images [47].

The PACE system is another multiple NDT techniques information system for data management and CAD coupling [48]. Such a system is used by an European electric power company for testing bottom head penetration tubes using four NDT methods: ultrasonic shear waves 45°, longitudinal waves, time-of-flight-diffraction (TOFD), and eddy currents (EC). This multi-modal system makes use of an enriched embedded TRAPPIST format.

Additional applications reported in the literature include surface texture measurement by combining signals from two sensing elements of a piezoelectric tactile sensor [49], and information fusion for improved condition monitoring [50]. The Indian Institute of Technology in Madras dealt with progressive tool wear monitoring through neural network based information fusion [51]. The potential of data fusion attracted the interest of several institutions from Asia. Some aspects of NDT data fusion for weld inspection were studied in detail at Kyoto University in Japan [52]. Hayashi *et al.* designed a system that allows simultaneous measurements of acoustic emission and computer tomography signals of a material under tensile load [53]. Although they do not combine information from both techniques in a synergistic manner, the obvious following step to their experiments would be to combine AE and CT data for inference purposes. Current research at the Honk Kong Polytechnic University aims at fusing radar images for improved inspection of buildings.

Rakocevic *et al.* used an array of 16 multiplexed ultrasonic probes for the inspection of newly rolled plates [54]. Their system can also deploy other types of sensors (e.g. UT, EC) hence making it suitable for data fusion and adequate for inspection in hostile environment (e.g. nuclear power station). Microstructure analysis is another area of application for which 3-D image reconstruction based on micro-structural volume reconstruction from a stack of serial sections is used to reveal information about tungsten grains [55].

With new multi-sensor techniques being developed - such as ultrasonic focusing with time reversal mirrors - the use of data fusion for industrial applications will increase as the amount of information generated will have to be processed in an efficient manner [56]. Forthcoming implementations resulting from current research and recently completed research will be to incorporate decision support information with data fusion for a complete automated, operator independent, inspection system. This will logically lead to further use of self navigation robotic systems for rapid larger area inspection, error-free and reliable, automated data acquisition and analysis.

1.4 PERSPECTIVES

More stringent regulations for structural integrity assessment to prevent accidents, protect the environment and citizens, as well as the need for better quality of manufacturing products in response to customers pressure will act as an incentive to develop data fusion systems further. The continuing need and desire to process more data more thoroughly will lead to an increase in the use of data fusion techniques to solve more complex problems. Not only will the results of non-destructive measurements be improved, the use and applications of data fusion to NDT will also lead to breakthroughs for the study and analysis of the microstructure of materials. It will become necessary for researchers and project managers to be familiar with the technique and to learn more about it.

One of the major limitations in applying data fusion techniques to the field of NDT and condition monitoring is that it requires multi-disciplinary expertise difficult to find in a single researcher. The full potential of NDT data fusion can only be appreciated by bringing together the work of technicians, engineers, mathematicians, physicists, and computer programmers. In brief, a team constituting members with complementary skills and diverse interests. The understanding and acceptance of this approach remain a difficult concept. But as Max Karl Ernst Ludwig Planck (1882-1944) once said, "A new scientific truth does not triumph by convincing its opponents and making them see the light, but rather because its opponents eventually die and a new generation grows up that is familiar with it."

REFERENCES

1. Gros X.E., *NDT Data Fusion*, Butterworth-Heinemann, 1997.
2. Dasarathy B.V., *Sensor fusion potential exploitation – Innovative architectures and illustrative applications*, Proc. IEEE, 1997, 85(1):24-38.
3. Hall D.L., Llinas J., *An introduction to multisensor data fusion*, Proc. IEEE, 1997, 85(1):6-23.
4. Spedding V., *Computer integrate disparate data*, Scientific Computing World, 1998, (38):22-23.
5. Georgel B., Lavayssière B., *Fusion de données: un nouveau concept en CND*, Proc. 6th European Conf. on NDT, 1994, Nice, France, 1:31-35.
6. Gros X.E. Strachan P., Lowden D.W., Edwards I., *NDT data fusion*, Proc. 6th European Conf. on NDT, 1994, Nice, France, 1:355-359.
7. Dessendre M., Thevenot F., Liot A., Tretout H., Exmelin A., *NDT and CAD data fusion for diagnosis assistance and diffusion bonding improvement*, Proc. COFREND Congress on NDT, 1997, Nantes, France.
8. Valet L., Mauris G., Bolon P., *A statistical overview of recent literature in information fusion*, Proc. 3rd Inter. Conf. on Information Fusion, 2000, Paris, France.

9. Nelson J.M., Cruikshank D., Galt S., *A flexible methodology for analysis of multiple-modality NDE data*, Proc. Review of Progress in QNDE, 1989, LaJolla, USA, 8:819-826.
10. Janney D.H., *Image processing in nondestructive testing*, Proc. 23rd Sagamore Army Materials Research Conf. on the Nondestructive Characterization of Materials, 1979, 409-420.
11. *Troll tunnel inspection ROV piloted in virtual reality mode*, Offshore, 1996, 56:141-142.
12. Gros X.E., Strachan P., Lowden D., *Fusion of multiprobe NDT data for ROV inspection*, Proc. MIS/IEEE Conf. Oceans '95, 1995, San Diego, USA, 3:2046-2050.
13. Gianaris N.J., Green R.E., *Statistical methods in material manufacturing and evaluation*, Materials Evaluation, 1999, 57(9):944-951.
14. Lallier E., Farooq M., *A real time pixel-level based image fusion via adaptive weight averaging*, Proc. 3rd Inter. Conf. on Information Fusion, 2000, Paris, France.
15. Petrović V., Xydeas C., *Computationally efficient pixel-level image fusion*, Proc. 3rd Inter. Conf. on Information Fusion, 2000, Paris, France.
16. Fu P., Hope A.D., Javed M.A., *Fuzzy classification of milling tool wear*, Insight, 1997, 39(8):553-557.
17. Goebel K., Badami V., Perera A., *Diagnostic information fusion for manufacturing processes*, Proc. 2nd Inter. Conf. on Information Fusion, FUSION'99, 1999, Sunnyvale, USA, 1:331-336.
18. Liu Z., Gros X.E., Tsukada K., Hanasaki K., *3D visualization of ultrasonic inspection data by using AVS*, Proc. 5th Far-East Conf. on NDT, 1999, Kenting, Taiwain, 549-554.
19. Pastorino M., *Inverse-scattering techniques for image reconstruction*, IEEE Instrumentation & Measurement Magazine, 1998, 1(4):20-25.
20. Summa D.A., Claytor T.N., Jones M.H., Schwab M.J., Hoyt S.C., *3-D Visualisation of x-ray and neutron computed tomography (CT) and full waveform ultrasonic (UT) data*, Proc. Review of Progress in QNDE, 1999, Snowbird, USA, 18A:927-934.
21. Vengrinovich V.V., Denkevich Y.B., Tillack G.R., Jacobsen C., *3D x-ray reconstruction from strongly incomplete noisy data*, Proc. Review of Progress in QNDE, 1999, Snowbird, USA, 18A:935-942.
22. Reed J.M., Hutchinson S., *Image fusion and subpixel parameter estimation for automated optical inspection of electronic component*, IEEE Trans. on Industrial Electronics, 1996, 43(3):346-354.
23. Bossi R.H., Nelson J., *NDE Data Fusion*, Proc. ASNT Fall Conf., 1997, Pittsburgh, USA.
24. Matuszewski B.J., Shark L.K., Varley M.R., Smith J.P., *Region-based wavelet fusion of ultrasonic, radiographic and shearographic non-destructive testing images*, Proc. 15th World Conf. on NDT, 2000, Rome, Italy, paper N°263.
25. Siegel M., Gunatilake P., *Enhanced remote visual inspection of aircraft skin*, Proc. Workshop on Intelligent NDE Sciences for Ageing and Futuristic Aircraft, 1997, El Paso, USA, 101-112.
26. Paul French, *Biomedical optics*, Physics World, 1999, 12(6):41-46.
27. Ault T., Siegel M.W., *Frameless patient registration using ultrasonic imaging*, J. of Image Guided Surgery, 1995, 1(2):94-102.
28. Strangman G., *Under doctors' orders for a digital revolution*, Scientific Computing World, 2000, (52):27-30.
29. Zachary J.M., Iyengar S.S., *Three dimensional data fusion for biomedical surface reconstruction*, Proc. 7th Fusion Conf., 1999, Sunnyvale, USA.
30. Gautier L., Taleb-Ahmed A., Rombaut M., Postaire J.-G., Leclet H., *Belief function in low level data fusion: application in MRI images of vertebra*, Proc. 3rd Inter. Conf. on Information Fusion, 2000, Paris, France.
31. Mangin J.F., Stévenet J.F., *Le cerveau s'affiche en dynamique et en haute définition*, CEA Technologies, 2000, (49):2.
32. Dupuis O., *Fusion entre les données ultrasonores et les images de radioscopie à haute résolution: application au contrôle de cordon de soudure*, PhD Thesis, 2000, INSA Lyon, France.
33. Dupuis O., Kaftandjian V., Drake S., Hansen A., *Ffreshex: a combined system for ultrasonic and X-ray inspection of welds*, Proc. 15th World Conf. on NDT, 2000, Rome, Italy, paper N°286.
34. Just V., Fleuet E., Gautier S., *The BE-MISTRAL project: a fully multi-technique approach from acquisition to data fusion*, Proc. 7th ECNDT, 1998, Copenhagen, Denmark 2:1608-1613.
35. Jain A.K., Dubuisson M.P., Madhukar M.S., *Multi-sensor fusion for nondestructive inspection of fiber reinforced composite materials*, Proc. 6th Tech. Conf. of the American Society for Composites,

1991, 941-950.
36. Yanowitz S.D., Bruckstein A.M., *A new method for image segmentation*, Computer Vision, Graphics and Image Processing, 1989, 46(4):82-95.
37. Barniv Y., Casasent D., *Multisensor image registration: experimental verification*, Proc. SPIE, Process. Images and Data from Optical Sensors, 1981, San Diego, USA, 292:160-171.
38. Akerman A., *Pyramidal techniques for multisensor fusion*, Proc. SPIE, Sensor Fusion V, 1992, Boston, USA, 1828:124-131.
39. Haberstroh R., Kadar I., *Multi-spectral data fusion using neural networks*, Proc. SPIE, Signal processing, sensor fusion, and target recognition II, 1993, Orlando, USA, 1955:65-75.
40. L. Ghouti, *A novel method for automatic defect classification using artificial neural networks*, Proc. Review of Progress in QNDE, 1999, Snowbird, USA, 18A:843-850.
41. Wang C., Cannon D.J., *Virtual-reality-based point-and-direct robotic inspection in manufacturing*, IEEE Trans. on Robotics and Automation, 1996, 12(4):516-531.
42. Hannah P., Starr A., Ball A., *Decisions in condition monitoring – an example for data fusion architecture*, Proc. 3rd Inter. Conf. on Information Fusion, 2000, Paris, France.
43. Stover J.A., Hall D.L., Gibson R.E., *A fuzzy-logic architecture for autonomous multisensor data fusion*, IEEE Trans. on Industrial Electronics, 1996, 43(3):403-410.
44. Russo F., *Recent advances in fuzzy techniques for image enhancement*, IEEE Instrumentation & Measurement Magazine, 1998, 1(4):29-32.
45. Benoist Ph., Besnard R., Bayon G., Boutaine JL., *CIVA poste d'expertise en contrôle non destructif*, Proc. 6th Europ. Conf. on NDT, 1994, Nice, France, 2:1311-1315.
46. Benoist P., Besnard R., Bayon G., Boutaine J.L., *CIVA Workstation for NDE: Mixing of NDE Techniques and Modeling*, Review of Progress in QNDE, Plenum Press, 1995, Snowmass Village, USA, 14B:2353-2360.
47. Nockemann C., Heine S., Johannsen K., Schumm A., Vailhen O., Nouailhas B., *Raising the reliability of NDE by combination and standardisation of NDT-data using the Trappist system*, Proc. Review of Progress in QNDE, 1996, Seattle, USA, 15A:1975-1982.
48. Just V., Gros P.O., Soors C., François D., *PACE a comprehensive multitechnique analysis system applied to the analysis of bottom head penetration tubes NDT data*, Proc. 7th Euro. Conf. on NDT, 1998, Copenhagen, Denmark, 2:1442-1447.
49. Dargahi J., Payandeh S., *Surface texture measurement by combining signals from two sensing elements of a piecoelectric tactile sensor*, Proc. SPIE AeroSense Conf., 1998, Orlando, USA, 3376:122-128.
50. Taylor O., Mayintyre J., *Adaptive local fusion systems for novelty detection and diagnostics in condition monitoring*, Proc. SPIE AeroSense Conf., 1998, Orlando, USA, 3376:210-218.
51. Mou J., Jones S.D., Furness R.J., *Sensor-fusion methodology for reducing product quality variation*, Proc. Computer Applications in Production and Engineering, 1997, London, U.K., 398-410.
52. Liu Z., *Studies on data fusion of nondestructive testing*, PhD Thesis, 2000, Kyoto University.
53. Hayashi K., Yamaji H., Nagata Y., Ishida T., *Combined method of high resolution x-ray computed tomography and acoustic emission for nondestructive testing*, Proc. Inter. Symp. on NDT & Stress-Strain Measurement FENDT, 1992, Tokyo, Japan, 361-366.
54. Rakocevic M., Wang X., Chen S., Khalid A., Sattar T., *Bridge B., Development of an automated mobile robot vehicle inspection system for NDT of large steel plates*, Insight, 1999, 41(6):376-382.
55. Tewari A., Gokhale A.M., *Application of three-dimensional digital image processing for reconstruction of microstructural volume from serial sections*, Materials Characterization, 2000, 44(3):259-269.
56. Fink M., Prada C., *Ultrasonic focusing with time reversal mirrors, Advances in Acoustic Microscopy*, Eds. A. Briggs, W. Arnold, Plenum Press, New York, USA, 1996, 2:219-251.

2 AN ROC APPROACH TO NDT DATA FUSION

Xavier E. GROS
Independent NDT Centre,
Bruges, France

2.1 INTRODUCTION

Receiver operating characteristic (ROC) analyses are usually carried out to test simple decision problems and hypotheses such as the presence or not of a defect. In this chapter, a case study is presented to show that a ROC performance evaluation approach can be implemented for non-binary decision problems in NDT. Two approaches are considered to evaluate the performance of fusing information produced by the examination of an impacted composite panel. The examination of this composite panel was performed using two different NDT techniques. The images generated with each technique were fused using image processing operations as described in chapter 3. Another fusion operation was performed statistically using a team ROC approach to combine decisions. In both cases, a ROC curve representing the fusion output was plotted and the performance of each fusion process evaluated based on these curves.

The ROC approaches discussed here make use of probabilistic inference theory. Probabilistic inference is well suited to reason under uncertainty, and is used to infer on the probability of events or hypotheses given additional information. Evidence about an event may take several forms; for example it could be experimental results from material testing or data from prior history of a structure. In addition, a data set would be necessary to determine the probability of having a defect in a structure (based on prior history, manufacturing process, lifetime); or the probability of a technique or instrument to detect and accurately size a specific defect (e.g. eddy current could be adequate to detect surface cracks but sizing may be poor due to the spatial resolution of the probe). The information about events and the dependencies among their outcomes can be expressed as conditional probabilities.

NDT signals can be associated with a hypothesis prior to decision. Confidence levels would have to be defined to efficiently perform a signal processing operation, and a mass function will be attributed to hypotheses (*cf* Dempster-Shafer theory). A level of confidence can be defined as the percentage of true-to-false defects in the region of the image studied [1].

Another aspect of probabilities in NDT is the use of probability of detection (POD) curves to assess the detectability of flaws with a specific NDT technique. Difficulties in POD result from the inability in preparing samples with simulated defects that are representatives of real defects. This approach is based on the principle of statistical detection theory in which one infers a POD and a probability of false alarm noted PFA (or probability of false call noted PFC) based on signal and noise distribution (figure 2.1) [1].

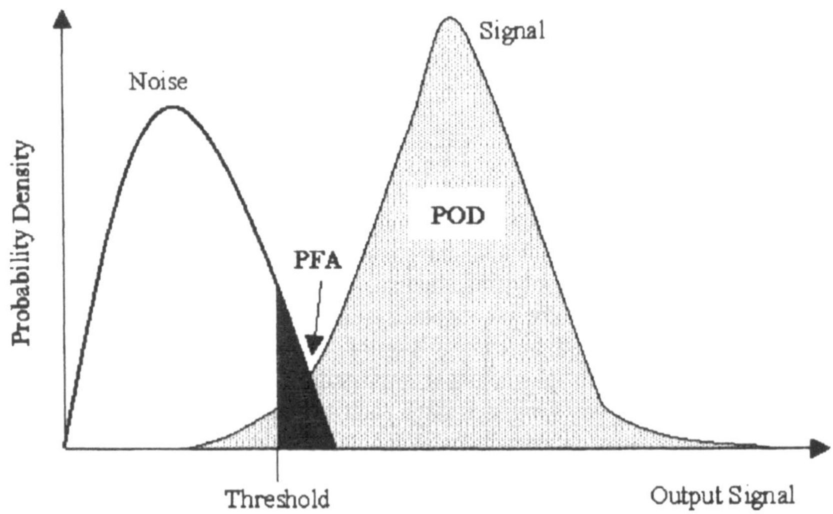

Fig. 2.1 Graphical representation of probability density *vs.* output signal (modified from [2])

From figure 2.1 it is possible to quantify the reliability of non-destructive inspection by plotting the POD against a defect size; the POD increasing with the size of the defect. A ROC curve is obtained by plotting the POD against the PFA. Both POD and ROC curves can be used to optimise an inspection procedure or to identify the parameters affecting an inspection (e.g. human factors, environmental conditions) or to evaluate operators' performance [2].

2.2 CASE STUDY

In this case study, data resulting from the inspection of impacted composite panels is analysed (see chapter 3). ROC curves have been established for each technique and for the fused data. Although this procedure can be carried out with any number of inspection techniques, only the case of eddy current testing and infrared thermographic testing are considered next.

2.2.1 Pixel Level Image Fusion

The probability of detection (POD) and probability of false call (PFC) were defined for each NDT technique based on experimental data. Such probability values can be gathered from repeated inspections or from historical data (i.e. data gathered from inspections of the same structure over a certain period). These were determined based on five categories, established as indicated in table 2.1, for defect sizing.

Table 2.1 Categories used to classify inspection results and establish probabilities

Categories	Correct Size		Incorrect Size	
	Eddy Current	Infrared Thermography	Eddy Current	Infrared Thermography
Correct Size (I)	10	7	2	3
Partially Correct Size (II)	8	6	2	3
Equivocal (III)	4	5	5	4
Partially Incorrect Size (IV)	2	4	7	7
Incorrect Size (V)	1	3	9	8

Two hypotheses have been tested based on the inspection results. The first hypothesis, noted H_0, is the hypothesis that the size of the defect (or damage) detected with eddy current is correct. The second hypothesis, noted H_1, is the hypothesis that the size of the defect (or damage) detected with infrared thermography is correct. In the following text the use of the terms 'true positive' and 'false positive' is preferred to POD and PFC. These results are summarised in table 2.2.

Table 2.2 True positive and false positive values for each technique and each category defined in table 2.1

	True Positive		False Positive	
	Eddy Current	Infrared Thermography	Eddy Current	Infrared Thermography
> I	0.92	0.88	0.60	0.72
> II	0.84	0.76	0.28	0.48
> III	0.64	0.60	0.12	0.28
> IV	0.36	0.32	0.04	0.12

Using the values from table 2.2, ROC curves can be plotted for each NDT technique (figure 2.2).

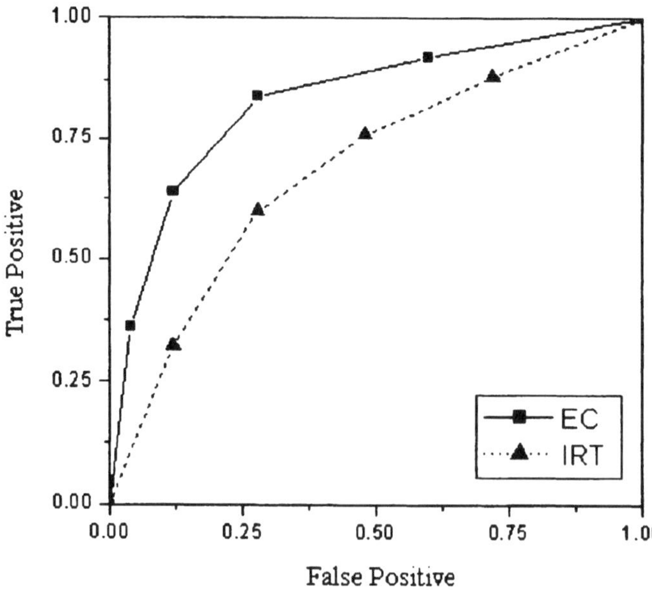

Fig. 2.2 ROC curves for eddy current (EC) and infrared thermography (IRT)

Image fusion at pixel level (*cf* chapter 3) led to a fused image that contained information from both NDT techniques. Multiple fused images can be obtained if inspection of the same structure is repeated several times, or if different fusion operations are performed. After each inspection, a pixel level fusion operation generates a new fused image. The size of the defect on the fuse images is estimated and compared to the 'real' defect size. Results of these operations are summarised in tables 2.3 and 2.4. For each fused image, the hypothesis that the size of the defect (or damage) detected on the fused image is correct is evaluated. The result of this evaluation is presented as a ROC curve in figure 2.3. These results are discussed in paragraph 2.2.3.

Table 2.3 Results in each category for the fused images

Category	Correct Size	Incorrect size
I	12	1
II	9	2
III	2	6
IV	1	7
V	1	9

Table 2.4 True positive and false positive values for the fused images

Category	True Positive	False Positive
> I	0.96	0.52
> II	0.88	0.16
> III	0.64	0.08
> IV	0.36	0.04

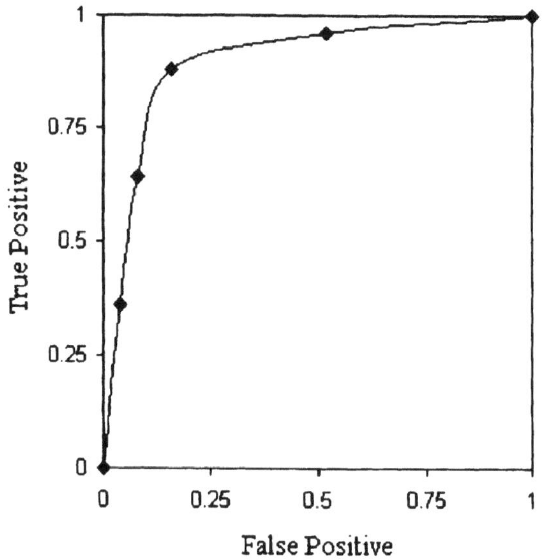

Fig. 2.3 ROC curve of the fused images resulting from pixel level fusion of EC and IRT images

2.2.2 A Team ROC Approach

In this section, a team ROC approach will be used to fuse individual decisions. A probability gives a measure of the degree of occurrence of a hypothesis. Among the various methods available the Bayesian approach for scientific inference takes into account raw data and prior knowledge that supplements the data (i.e. it assigns probability to theories and evidence). With Bayes' theory one who analyses two data sets collectively or sequentially, should arrive at the same answer. Probabilities are assigned rather than measured because they describe the plausibility of propositions and reflect our knowledge about the 'truth' of such propositions. Nevertheless, in certain cases a probability can be measured as it is a frequency. Probability assignment is difficult, especially if little is known of the material tested or the technique used. Since Bayesian statistics relies on conditional states of knowledge, it is necessary to explicitly identify an appropriate prior distribution that describes what prior knowledge (or ignorance) is available for a hypothesis. The prior distribution can be used to express any desired prior state of knowledge; ranging from total ignorance (called 'non-informative priors' or 'least informative priors') to limited information ('maximum entropy priors') to quite informative knowledge ('conjugate priors') [3]. Non-informative priors may rise for analysis of two exclusive propositions that do not show an *a priori* preference for any particular value of the parameter.

Equal probability is given by the prior distribution to each possible value of the parameter. Bayes' theorem states that:

$$p(x \mid y,i) = \left[\frac{p(y \mid x,i)p(x \mid i)}{p(y \mid i)}\right] \tag{2.1}$$

Where p(x|y,i) is known as the *posterior* probability; which is the probability of hypotheses *x* and *y* based on the information i.

Let's consider two hypotheses noted H_0 and H_1. H_0 is the hypothesis that the size of the defect detected with eddy current is correct, and H_1 the hypothesis that the size of the defect detected with infrared thermography is correct. A value can be associated with each hypothesis that defines the probability of occurrence of one hypothesis. For example, the probability of obtaining H_0 is noted $p(H_0)$ (or P_0) and is equal to 0.7. Therefore, $p(H_1) = 1 - p(H_0) = 1 - P_0 = P_1 = 0.3$. If no information is available, then $p(H_0^{'}) = p(H_1^{'}) = 0.5$.

The experimental result from a non-destructive examination can be used to make a decision towards the presence of a defect or its dimensions. The output of a NDT system can be defined as a decision maker (as the operator will make a decision based on signal assessment). Each decision maker forms a local binary decision about a hypothesis that can be noted $u_i \in U_i = \{0,1\}$, where $u_i = j$ indicates that a decision maker *i* tends to support hypothesis H_j ($j = 0, 1$). Pete *et al.* defines the optimal aggregation rule for the fusion centre as $\gamma \in \Gamma$ (where Γ is the set of all possible rules) such that the expected cost $J(\gamma)$ determined by equation (2.2) is minimised [4]:

$$\begin{aligned} J(\gamma) &= E\{C(u_0, H_j)\} = \sum_{j=0}^{1} \sum_{u_0=0}^{1} C(u_0, H_j)P(u_0, H_j) \\ &= \sum_{j=0}^{1} \sum_{u_0=0}^{1} C(u_0, H_j)P(u_0 \mid H_j)P(H_j) \end{aligned} \tag{2.2}$$

where $C(u_0, H_j)$ is a cost value associated with (u_0, H_j), and representing the cost of a particular decision given the true event H_j. After further manipulations, it is possible to write equation (2.2) as follows:

$$\begin{aligned} J(\gamma) &= P_0C(0, H_0) - P_FP_0C(0, H_0) + P_FP_0C(1, H_0) + P_1C(0, H_1) - P_DP_1C(0, H_1) \\ &\quad + P_DP_1C(1, H_1) \\ &= P_F[P_0C(1, H_0) - P_0C(0, H_0)] - P_D[P_1C(0, H_1) - P_1C(1, H_1)] + [P_0C(0, H_0) \\ &\quad + P_1C(0, H_1)] \end{aligned}$$

Which can be written in its simplest form as:

$$J(\gamma) = aP_F - bP_D + c \tag{2.3}$$

where, for one specific NDT technique, $P_D = p(S1/H_1)_{NDT1}$ and $P_F = p(S1/H_0)_{NDT1}$ (similarly $P_D = p(S2/H_1)_{NDT2}$ and $P_F = p(S2/H_0)_{NDT2}$ for the other NDT technique). Let's note $P_D(\lambda_j)$ and $P_F(\lambda_j)$, (for j = 1, 2), the detection and false alarm probabilities when λ_i is the decision threshold. It is then possible to mathematically express the fused detection and false alarm probabilities, respectively P_D' and P_F':

$$P_D' = \pi P_D(\lambda_1) + (1-\pi)P_D(\lambda_2)$$
and
$$P_F' = \pi P_F(\lambda_1) + (1-\pi)P_F(\lambda_2) \tag{2.4}$$

Which after calculations for given cost values gives:

$$P_D' = 0.65P_D(\lambda_1) + 0.35P_D(\lambda_2)$$
and
$$P_F' = 0.65P_F(\lambda_1) + 0.35P_F(\lambda_2) \tag{2.5}$$

The values of $P_D(\lambda_j)$ and $P_F(\lambda_j)$ are graphically determined, and a team ROC joint combination set of values (P_F', P_D') can be calculated (table 2.5). For a more complete description of the mathematics involved it is best to refer to Pete *et al.* [4]. A team ROC graph can be plotted using the values from table 2.5 that represent the joint fusion process of the individual decisions from each NDT technique used (i.e. eddy current and thermography) (figure 2.4).

Table 2.5 Team ROC joint combination

P_D' True Positive	P_F' False Positive
0.906	0.642
0.812	0.350
0.626	0.176
0.346	0.068

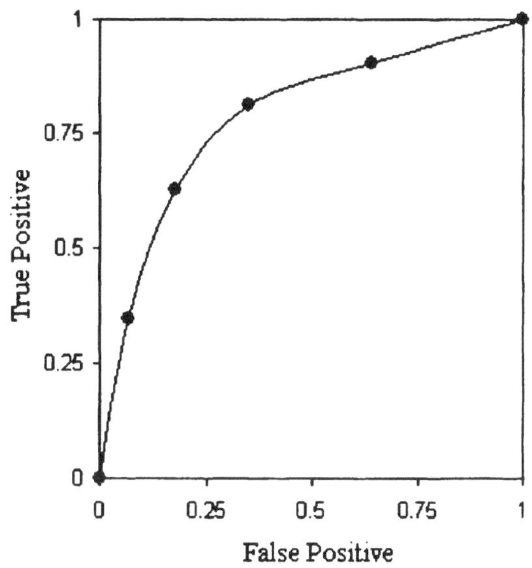

Fig. 2.4 ROC curve of the fused decisions from eddy current and infrared testing

2.2.3 ROC Performance Evaluation

In order to select the fusion process best suited for an application, it is necessary to evaluate the outcome of a fusion operation. One way to estimate the performance of a data fusion process is to calculate the *K* value of ROC curves prior and after fusion. The magnitude of vector *K* relative to the diagonal of the ROC curve gives an estimate of the performance of an event. If the *K* value of the fused ROC curve is greater than any individual ROC curve for each NDT technique, then an improvement has been achieved and the fusion process may be worth considering. In the event of a *K* value lower than the individual *K* value of each technique prior to fusion, it can be concluded that the fusion process did not improve the inspection performance. The operator would have to decide if the fusion algorithm is adequate or if a different algorithm needs to be developed (or even if fusion is really necessary for this application). The performance parameter *K* is calculated based on [5]:

$$K = 1 - 2\sqrt{(1 - \text{POD}) \times \text{PFC}} \quad (2.6)$$

The POD and PFC values (or true and false positive values) are graphically determined at the tangential point on a ROC curve as indicated in figure 2.5.

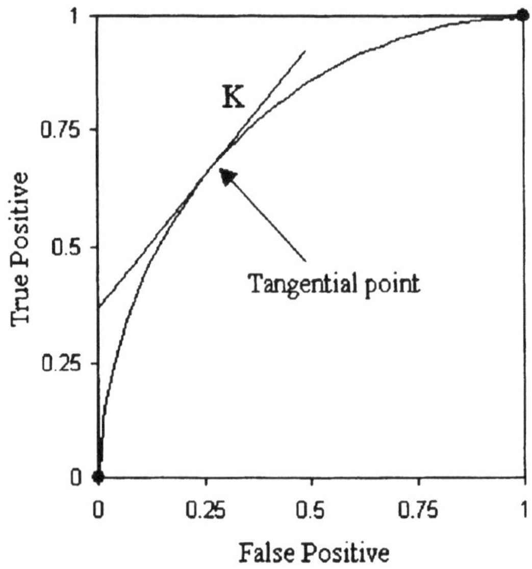

Fig. 2.5 ROC curve for calculation of the performance parameter *K*

The *K* values determined for each NDT technique as well as for the fused output from the pixel level data fusion operation and the team ROC approach are summarised in table 2.6.

Table 2.6 Values of performance parameter *K* for EC and IRT, and their fused image and decision outputs

NDT Technique	K values
Eddy Current	0.584
Infrared Thermography	0.331
Fused EC + IRT (pixel level)	0.661
Fused EC + IRT (team ROC)	0.487

From the above values, one may notice that the performance for defect sizing based on evaluation of fused images at pixel level is greater than of the team ROC approach. This can be explained by the fact that a human operator will estimate the size of a defect once image fusion has been completed. Hence, human factors such as experience and material knowledge, will affect the final decision. Those parameters are difficult to model as they are specific to each operator, and vary from one inspection to the next. Additionally, several image fusion algorithms have been tested and the false and true positive values were determined for the whole processes. With the team ROC approach, decisions from each technique are fused at a decision level (i.e. image fusion operation does not occur). The main advantage of the team ROC approach is that it takes into consideration the advantages and limitations of each technique, thus providing an optimised decision output. In any case, one may also notice an improvement in the fused ROC curves compared with the original infrared thermographic ROC curve. In the case of pixel image fusion, a performance greater than any individual ROC curves is achieved. Several other approaches could be considered to fuse information but in any case there is no optimal fusion rule. For example, possibility theory could be used to handle uncertain information because it has the ability to merge data from different sources to increase the overall quality of a NDI. The possibility theory is best suited to deal with contradictory data and to take account of vagueness or uncertainty of variables [6].

2.3 CONCLUSION

In this chapter, it was demonstrated that the combination of information at a statistical and probabilistic level actually increases knowledge about defect detection or sizing, and helps decision making. The team ROC approach presents the advantages of not putting too much faith in one NDT technique only (or one decision) and considers all potential decisions based on the information provided by each NDT system. Although examples were given for only two NDT techniques, the same operations can be performed for a greater number of systems. Indeed, since the fusion reliability increases with the amount of information incorporated in the decision, the use of more than two techniques is recommended [7]. The main interest resides in the fact that ROC curves can be used to graphically represent and quantify the performance of a fusion operation. Because "sensor data fusion is now an established approach to increase the performance of a profusion of sensing systems", the use of ROC curves should also become more widespread to evaluate fusion procedures [8]. It will also facilitate performance comparison of fusion processes by providing a common reference criteria.

REFERENCES

1. Thompson R.B., *Overview of the ETC POD methodology*, Proc. Review of Progress in QNDE, 1999, Snowbird, USA, 18B:2295-2304.
2. Wall M., Wedgwood F.A., Burch S., *Modelling of NDT reliability (POD) and applying corrections for human factors*, Proc. 7th ECNDT, 1998, Copenhagen, Denmark, 2:2108-2115.
3. Graps A., *Probability offers link between theory and reliability*, Scientific Computing World, 1998, (42):25-27.
4. Pete A., Pattipati K.R., Kleinman D.L., *Methods for fusion of individual decisions*, Proc. IEEE American Control Conf., 1991, Boston, USA, 2580-2585.

5. Boogaard J., Dijk G.M., *NDT reliability and product quality*, NDT&E Inter., 1993, 26(3):149-155.
6. Delmotte F., Borne P., *Modeling of reliability with possibility theory*, IEEE Trans. on Systems, Man and Cybernetics, Part A: Systems and Humans, 1998, 28(1):78-88.
7. Horn D., Mayo W.R., *NDE reliability gains from combining eddy-current and ultrasonic testing*, NDT&E Inter., 2000, 33(6):351-362.
8. O'Brien J., *An algorithm for the fusion of correlated probabilities*, Proc. Fusion'98, 1st Inter. Conf. on Multisource-Multisensor Information Fusion, 1998, Las Vegas, USA, 565-571.

3 PIXEL LEVEL NDT DATA FUSION

Xavier E. GROS
Independent NDT Centre,
Bruges, France

3.1 INTRODUCTION

Pixel level data fusion consists of combining images from multiple NDT systems with different physical properties, to reveal complementary or redundant information about the physical and mechanical characteristics of a material. Multiple sensory output information exists that can be in the form of a one-dimensional signal such as an analogue display on an oscilloscope, 2-D arrays such as images or radiation fields, or 3-D arrays if one is dealing with volumetric images. Image fusion can be defined as the combination of multiple images to generate a single output that will contain information extracted during the fusion of these images. It could be a combination of 2-D images to generate another with an improved signal-to-noise (S/N) ratio, or revealing details that could have been hidden when looking at each image separately. It could also be the combination of multiple 2-D images to generate one with additional dimensionality (i.e. 3-D). Pixel level data fusion evolved from simple image fusion operations (e.g. AND, OR operations), to pyramid decomposition based fusion, and to wavelet transform based fusion.

The fusion of multiple infrared and visible light images is common practise for target location and recognition [1], but pixel level fusion is also becoming an accepted technique for NDT applications. For example, Reed and Hutchinson performed image fusion with subpixel edge detection, and parameter estimation was used for automated optical inspection of electronic components [2]. Dromigny *et al.* used a Bayesian based data fusion approach to improve the dynamic range of real time X-ray radiographs by integrating information from the same real time X-ray imaging system gathered under two different acquisition conditions [3]. Mina *et al.* combined real and imaginary components from eddy current images [4]. Each image is transformed prior to fusion using orthogonal transformations (e.g. FFT, wavelet transform, discrete cosine transform) with Matlab software. The major features from each image is kept during the fusion operation but it was observed that an improvement may result when the quality of the original images is poor, while the fusion of images with identical good qualities may not result in an improvement. Q-transforms to the diffusion domain have also been used to address registration issues in the fusion of eddy current and ultrasonic data [5]. The transformation allows the superposition of the transform field on the eddy current field. Barniv and Casasent showed that the pixel correlation coefficient provides an adequate measure of the statistical independence of grey levels in a multisensor image pair [6].

Overall, the reliability of a non-destructive examination (NDE) can be improved through the combination of redundant information, as well as through the fusion of complementary information. The aim of image fusion in NDT is to provide the inspector with additional information about the structural integrity of a material, enhance the interpretation of NDT images, provide extra defect dimensionality, provide reliable information about defect location and size, and reduce noise. For example, the detection of local inhomogeneities in composites with X-radiography, ultrasonic C-scan and acoustic emission (AE) was performed by Jain *et al.* [7]. They showed that complementary information from multiple techniques helped reduce noise and improve defect detection. While AE peaks show defect location, image segmentation is used to extract defect areas from radiographs and C-scans. A fusion technique reconstructs a complete map that gives location and shape of defects. Common problems include registration or spatial alignment of multiple images, and integration or fusion of information extracted from multiple images. Image segmentation for defect extraction from NDT images with poor contrast has been performed using the Yanowitz and Bruckstein's method to construct a thresholding surface by interpolating the image grey levels at points where the gradient is high, indicating probable object edges [8]. This method consists of assigning a threshold value to each pixel.

A weight reduction combined with high stress resistance made composite materials an attractive alternative to more conventional materials such as aluminium alloys, especially for aerospatial and aeronautics applications. But as advanced materials are being developed, there is limited understanding of their mechanical behaviour during use and of damage propagation. In addition, the anisotropy and heterogeneousity of these new materials make their NDE a difficult task with conventional non-destructive techniques. In order to guarantee safety and to meet more stringent regulations, existing NDT techniques had to be adapted or new ones developed for in-service inspections and quality testing at the manufacturing level. An increase in the use of composite materials for primary structures will lead to the development of more accurate and efficient non-destructive inspection methods. Impact damage and delamination are among the most common type of defects in composites. These may occur at the manufacturing level as well as during in-service applications. It is now common knowledge that no single non-destructive method can accurately detect and quantify impact damage in composite materials. The detection of low energy impact damage remains difficult, even with state-of-the-art ultrasonic apparatus. Probably the most widely used NDT technique for inspection of composites is ultrasound [9]; however, despite its popularity, there are no widely accepted standards or procedures for ultrasonic testing of composites [10]. In addition, ultrasonic testing of composites requires the use of a water couplant, the inspection is often complex and defect sizing can be difficult with new materials due in part to high attenuation and wave scattering [11]. In order to overcome some of the limitations of ultrasonic testing of composites (e.g. need of a couplant), non-contact methods such as infrared thermography and shearography have been developed by industry. Still, these methods remain costly, and a need for an efficient and low cost technique for composites testing remains. Previous research work showed that eddy current is a potential alternative to detect and quantify low energy impact in carbon fibre reinforced plastics (CFRPs) [12-13].

Despite their advantages and limitations, each NDT technique can only provide limited information about a flaw or a defect. Indeed, the different physical properties measured with each technique reveal only a specific feature of the structure tested. In the case of impact damage, infrared thermography may reveal impact location but provides very little detail regarding the extent of the damage. On the other hand, eddy current testing can reveal sub-surface delaminations and broken fibres. Following defect detection with one method, the use of another is often required to check the validity of the first one and to gather additional information about the size and dimension of a defect. This extra information may help fracture engineers assess the remaining life of a structure. Nevertheless, the use of multiple NDT techniques is both costly and time consuming. It also generates a large volume of data that will have to be processed rapidly in a meaningful way. If improperly processed NDT signal interpretation can become a complex task, even for a trained operator. Analogue signal displays may sometimes be noisy and thus difficult to interpret. Although signal processing techniques are now common place, computer graphics visualisation facilities are additional tools for rapid defect detection and characterisation. Visual data analysis through image displaying the inspected area, and any defect areas, is often easier to interpret than an analogue signal on an oscilloscope [14]. Still, each image may convey only limited information about a defect. A more accurate vision of the damaged area could be obtained by combining images from more than one NDT technique.

The application of data fusion processes to combine NDT images at pixel level may help reduce uncertainty and errors of interpretation, and thus allows a more accurate evaluation of the structural integrity of a material [15]. Depending on the input images and on the data fusion architecture and process used, various types of information can be extracted. The combination of images can be performed using arithmetic, probabilistic and statistical algorithms such as ensemble averaging, Bayes or Dempster-Shafer theories. More recent theories based on wavelet analysis can also help in reducing ambiguity that may occur due to incomplete or missing attributes.

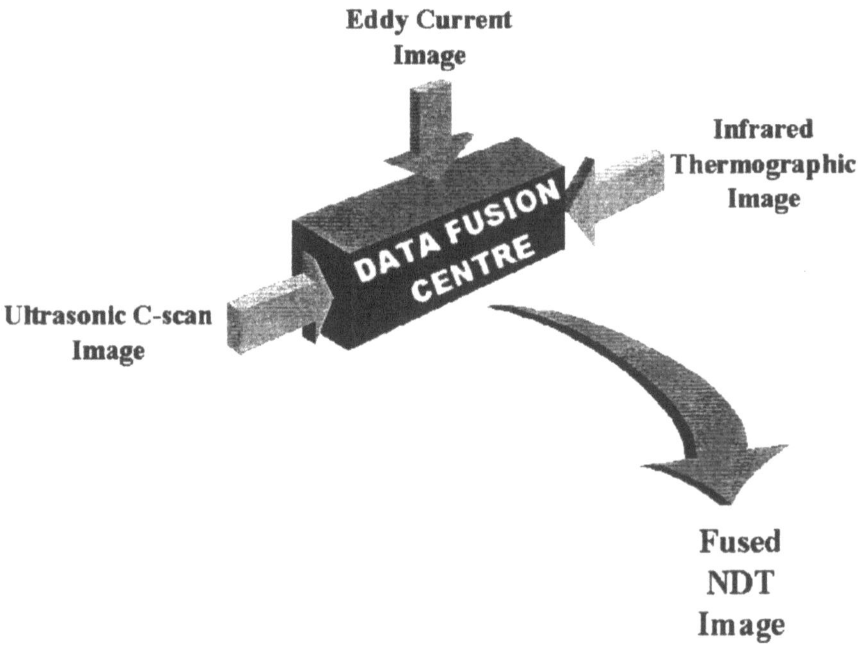

Fig. 3.1 Schematic of a pixel level image fusion process

The problem of combining images resulting from the non-destructive examination of an impacted CFRP composite panel is addressed in this chapter. In order to provide more comprehensive information of the damage extent, images from eddy current, infrared thermographic and ultrasonic inspections were fused to generate a single image (figure 3.1). The output signals from these NDT techniques were combined (table 3.1). Experimental results are presented, and the efficiency of each fusion technique discussed.

Table 3.1 Description of the output format and information obtained with the techniques used for the inspection of an impacted composite panel

NDT Technique	Signal output Format	Information gathered
Infrared Thermography	Thermograph (direct 2-D image)	Defect location Rough estimate of defect size (surface + subsurface)
Eddy Current	Analogue signal converted to a map of the inspected area (artificially generated 2-D image)	Defect location Defect size (surface + subsurface) Damage extent
Ultrasonic	Analogue signal and image map (C-scan 2-D image)	Defect location Estimate of defect size (subsurface + internal) Estimate of damage extent

3.2 INSTRUMENTATION AND EXPERIMENTAL SET-UP

A hybrid fabrics PA66 CFRP composite panel similar to the one used in industrial applications and manufactured by Nittobo in Japan was damaged by a 2 Joule impact. The specimen, a 150 × 150 × 2 mm plain weave carbon fibre composite panel, was impacted using a computer controlled drop tower mechanism. The damaged material was inspected with eddy current, infrared thermography and ultrasound. The eddy current inspection was performed using a commercial system, namely the Hocking Phasec 3.4D, by scanning the surface of the material with a broad band 500 kHz - 4 MHz unshielded surface probe (reference 130P1 and balance load reference 5A001) and recording the signal output on a computer. Computer visualisation of the area inspected was generated at a later stage. Figure 3.2 shows the extent of damage caused by the low energy impact as detected with eddy current. Next, an infrared thermographic examination was carried out after pre-heating the sample up to 100°C. A Nikon Laird 3A camera was used to monitor variations of temperature as the specimen cools down. Thermographs were recorded on a computer in real time. Figure 3.3 is the thermograph showing impact damage as detected with the infrared camera.

Ultrasonic C-scan examinations were carried out using two different instruments: an ultrasonic C-scan system from Krautkramer Japan, and the AT7500 type scanning acoustic tomograph from Hitachi. Results from the inspection performed with the Hitachi system at a frequency of 10 MHz are shown in figure 3.4. The inspection was computer controlled using an automated high-speed scanner (300 mm/s) and signal output was displayed on a monitor with a resolution of 600 × 400 pixels. The pulse echo mode, in which the same transducer emits and receives ultrasonic waves, was used. This technique has the advantage that access to only one side of the structure inspected is required. The pulse echo technique is effective on the near side skin laminate only and the sensitivity decreases as a function of depth. Results of the ultrasonic examination performed with the ultrasonic C-scan system from Krautkramer at a frequency of 25 MHz are presented in figure 3.5.

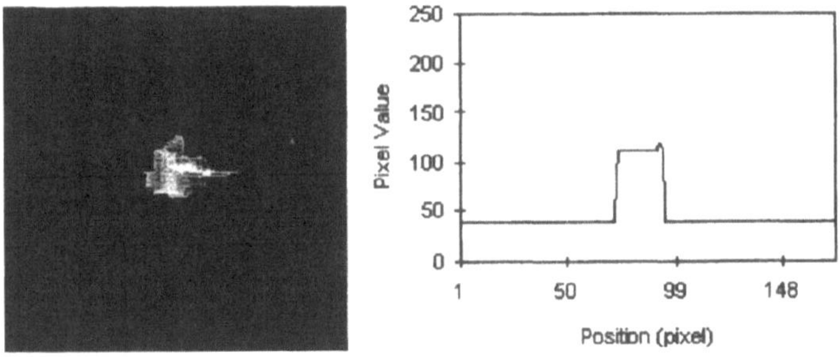

Fig. 3.2 Result of the eddy current inspection with pixel value profile along the centreline of the image

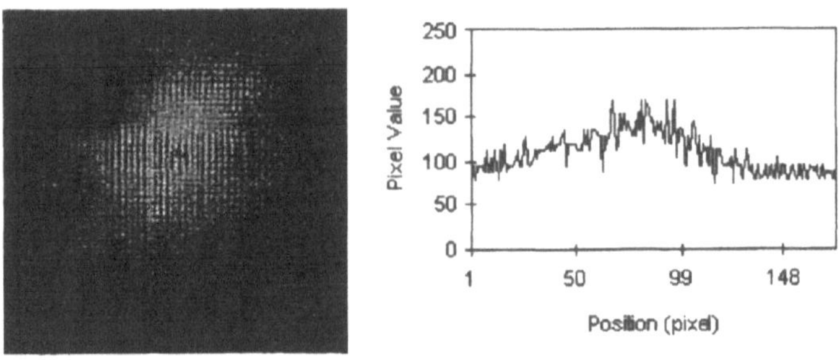

Fig. 3.3 Result of the infrared thermographic inspection with pixel value profile along the centreline of the image

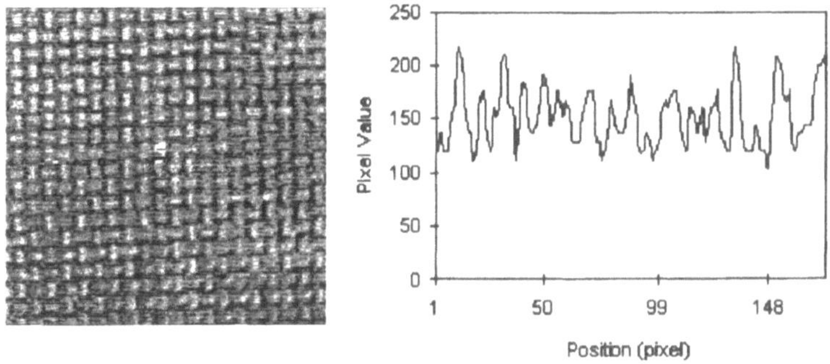

Fig. 3.4 Result of the ultrasonic inspection performed with Hitachi AT7500 system with pixel value profile along the centreline of the image

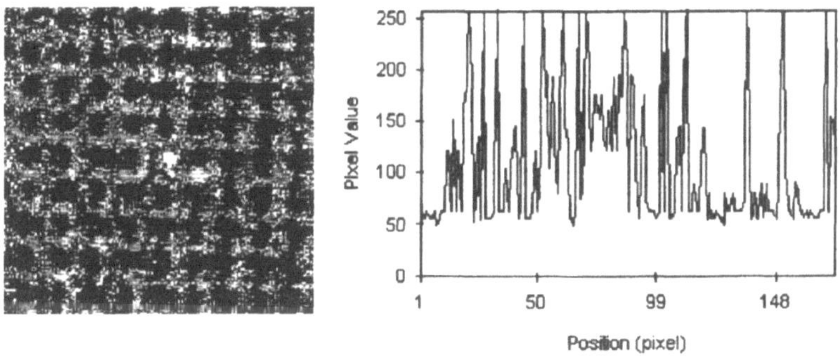

Fig. 3.5 Result of the ultrasonic inspection performed with Krautkramer ultrasonic C-scan system with pixel value profile along the centreline of the image

These inspection results clearly illustrate discrepancy that may arise when trying to quantify a defect. Indeed, the difference between each image is relatively obvious. Each figure is accompanied by a pixel amplitude profile calculated along the centreline of the image which can be used to provide an estimate of the efficiency of a data fusion process.

3.3 PIXEL LEVEL NDT DATA FUSION AND IMAGE PROCESSING

Data visualisation is a pre-processing step which benefits from advanced visualisation techniques. Computer graphics visualisation of NDT data allows the extraction of useful information form large data sets and visual data analysis can help signal interpretation [15]. However, information conveyed on an image may be noisy or may only be partial. Poor signal-to-noise ratio, pixel limitation or grain sizes on radiographic films are factors which may affect the quality of an image. The inspection results obtained with either eddy current, ultrasonic or infrared thermography have been transformed into a common format. Data alignment was performed such that each image represents the same view of the specimen. Aligning each image with the scanned area solved the registration problem. Therefore, the pixel value on each image relates to the same position on the sample. After establishing a common spatial reference the image is transformed to obtain the same dynamic range; in this case 256 grey levels. The damaged area as detected with eddy currents is shown as a grey scale image in figure 3.2. Although one may have a general idea of the location of the damage, it may be difficult to locate the exact boundaries and thus its full extent. In order to determine the damaged area more accurately, contours were traced on the image or pixel amplitude profile plotted. Figure 3.6a shows the result of this contour operation; one can distinguish the damaged area more easily. It is difficult to clearly identify the impact damage on the infrared thermographic image (figure 3.3) due in part to a lack in resolution depth. Although the thermographic camera was sensitive enough to distinguish the weave pattern of the carbon fibres, the impact damage was detected more difficulty and appears as a small indent at the centre of the image. As in the previous case, contours were traced to facilitate defect location (figure 3.6b) but in this case no improvement was achieved. Prior to fusion, further image processing was performed using the wavelet theory [16].

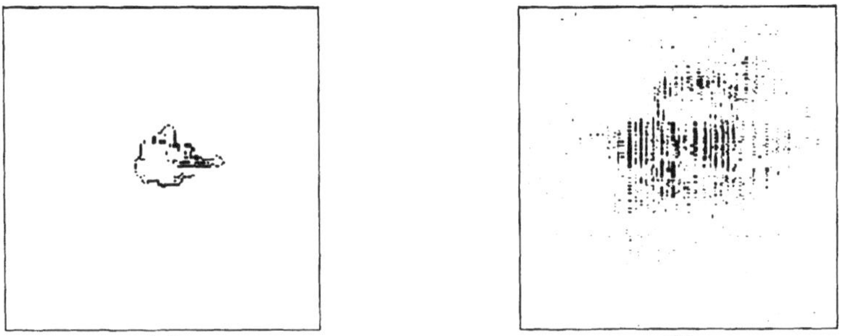

(a) (b)

Fig. 3.6 Results of the contour operation performed in fig. 3.2 (a) and fig. 3.3 (b)

Pixel level fusion of eddy current and infrared thermographic images, as well as the fusion of images from three NDT techniques (i.e. eddy current, infrared thermography and ultrasound), was performed using different fusion processes. The fusion operation was performed using feature based inference techniques as summarised in figure 3.7. Eddy current, thermographic, and at a later stage ultrasonic images were used as input files to the data fusion centre in which the fusion process took place. Conventional probabilities, Bayesian analysis [17], Dempster-Shafer theory [18] or a fuzzy logic approach [19] are among the most commonly used theories.

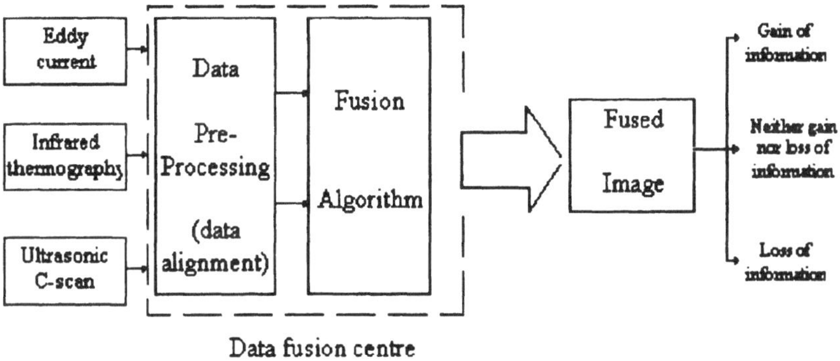

Fig. 3.7 Schematic of the architecture of a data fusion system and possible outcomes

Validation or rejection of a data fusion process is determined by analysing whether information was gained or lost. Results obtained by combining the NDT images on a pixel-by-pixel basis using conventional data fusion processes are presented in the following section.

3.3.1 Maximum Amplitude

A comparison of pixel values in view of a selection of the maximum amplitude may help in extracting valuable information from each image. Figures 3.8 and 3.9 show the result of the maximum amplitude operation performed on the eddy current, infrared thermographic images and the two ultrasonic images. In figure 3.8, one can notice that the defect area detected with eddy currents is reduced (although the shape of the damaged area remains similar), and that the impact detected with thermography remains visible. Figure 3.9 conveys information from all images, with the weave structure as detected with the Hitachi system remaining visible while the image from Krautkramer ultrasonic C-scan inspection has been largely soften. Still, the defect area detected with all techniques has been combined and seems clearer on the fused image, thus facilitating defect detection.

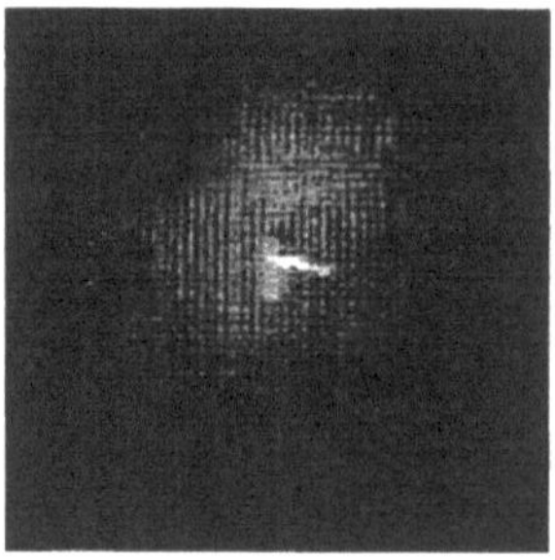

Fig. 3.8 Result of the maximum amplitude operation performed on eddy current and infrared images

Fig. 3.9 Result of the maximum amplitude operation on eddy current, infrared and two ultrasonic images

3.3.2 AND Operation

Probably the simplest form of pixel level data fusion is to use an AND operator. Figure 3.10 is the result of the integration on the four images. Fibre weave is no longer visible and the damaged area may not be so distinct, but by plotting the pixel amplitude value along the centreline of the image, a peak clearly stands out marking the position of the impact.

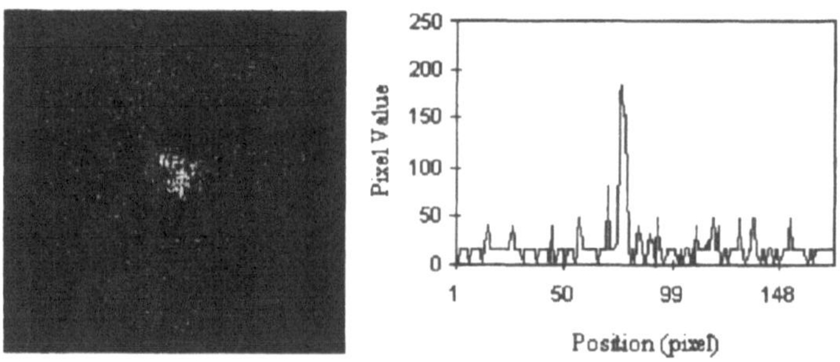

Fig. 3.10 Result of the AND operation performed on four NDT images, with pixel amplitude plot along the centreline of the image

3.3.3 Averaging

Ensemble averaging can be mathematically expressed as [20]:

$$\bar{x} = \frac{\sum\limits_{i=1 \to n} x_i}{n} \tag{3.1}$$

The result of ensemble averaging at pixel level on the four NDT images is shown in figure 3.11. A slight noise reduction was obtained but confirmation of the location of the defect from the superposition of information from all techniques appears more difficult than in figure 3.10.

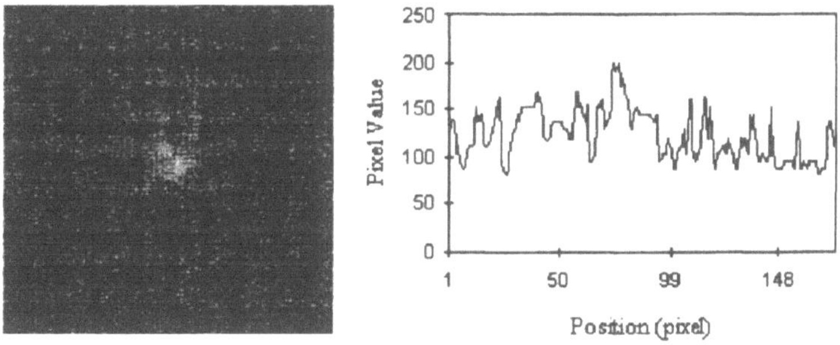

Fig. 3.11 Result of ensemble averaging performed on four NDT images, and pixel amplitude plot along the centreline of the image

3.3.4 Weighted Averaging

In the event where more knowledge or confidence towards one NDT technique is available, weighted averaging is a possible option for data fusion [20]. It consists of associating a value with a sensor based on previous inspection results used to establish POD curves. This value is referred to as weight, as it tends to associate a level of confidence on the output result of a NDT system. Equation 3.2 was used to compute weight-averaged images resulting from the combination of eddy current, infrared and ultrasonic images.

$$W\bar{x} = \frac{\sum\limits_{i=1 \to n} w_i x_i}{\sum\limits_{i=1 \to n} w_i} \tag{3.2}$$

Figure 3.12 illustrates the fused image resulting from the weighted average by associating a weight of 0.7 with the eddy current pixel values (w_{ec} = 0.7), a weight of 0.3 was associated with the infrared thermographic image (w_{ir} = 0.3), and a weight of 0.4 to the Hitachi ultrasonic image (w_{ut} = 0.4). Figure 3.13 shows the fused image obtained with a weight of 0.8 to the eddy current pixel values (w_{ec} = 0.8), a weight of 0.2 to the infrared thermographic image (w_{ir} = 0.2), and a weight of 0.2 to the Hitachi ultrasonic image (w_{ut} = 0.2). It can be seen that as the weight increases, information from the image with the lower weight is reduced, and the woven structure of the composite that clearly appeared on the ultrasonic image is almost totally removed. Additionally, an improvement in S/N ratio can also be noticed.

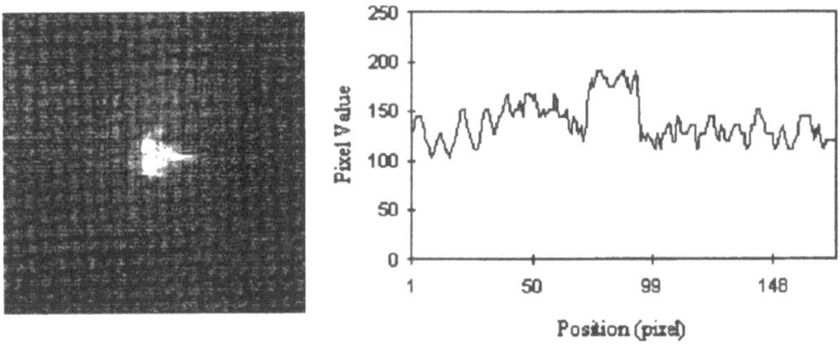

Fig. 3.12 Result of weighted averaging performed on the eddy current (w_{ec} = 0.7), infrared (w_{ir} = 0.3), and Hitachi ultrasonic (w_{ut} = 0.4) images, with pixel amplitude plot along the centreline of the image

3.3.5 Bayesian Analysis

Pixel level data fusion using the Bayesian theory was carried out by combining conditional and *a priori* information gathered from each NDT technique used. The result of such an operation is presented in the form of an *a posteriori* distribution. Prior to applying Bayes's theory, one has to define the probability of having a specific pixel value *pv* given a defect *d*. Such probabilities for the eddy current, infrared, Hitachi and Krautkramer ultrasonic images are noted $p(pv/d)_{ec}$, $p(pv/d)_{ir}$, $p(pv/d)_{uth}$ and $p(pv/d)_{utk}$ respectively. Baye's rule can be written as [17]:

$$p(d/pv) = \frac{p(pv/d)p(d)}{p(pv)} \tag{3.3}$$

with $p(pv) = \sum_i p(d_i) p(pv / d_i)$

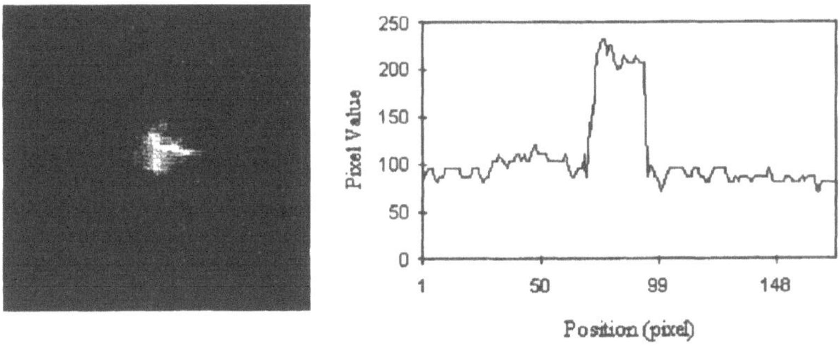

Fig. 3.13 Result of weighted averaging performed on the eddy current ($w_{ec} = 0.8$), infrared ($w_{ir} = 0.2$), and Hitachi ultrasonic ($w_{ut} = 0.2$) images, with pixel amplitude plot along the centreline of the image

The general form of equation 3.3 is:

$$p(d_i / pv) = \frac{p(pv / d) p(d_i)}{\sum_i p(d_i) p(pv / d_i)} \tag{3.4}$$

where p(d) is the prior probability of having a defect *d* in the specimen inspected, and p(d/pv) is known as the *posterior* probability of hypothesis *d* knowing the pixel value. Additional details about the Bayesian theory can be found in the literature [15,17] and are thus not incorporated in this chapter. If no prior knowledge of the specimen and the type of defect which may occur is available, estimating the value of p(d) can be difficult. In such cases it is common practice to assume a 50% probability of defect occurrence. Similarly, it can be difficult to determine conditional probabilities. It was noted from previous experiments [15], that eddy currents are more accurate in determining defect damage in composites compared to infrared thermography. So, it is safe to assume that the conditional probability associated with the eddy current inspection will be higher than the one associated with infrared thermography. Figure 3.14 shows fused images obtained from the fusion of the eddy current and infrared images for increasing values of p(d), and for $p(pv / d)_{ec} = 0.7$ and $p(pv / d)_{ir} = 0.5$. In the case where the value of p(d) is equal to 0.1, the resulting image conveys very little detail as it is assumed that the probability of having a defect is extremely low (figure 3.14a). Figure 3.14b is the pixel amplitude plot of figure 3.14a and is used to facilitate signal interpretation and to try to identify any potential defect area. Indeed, figure 3.14a appears very dark and one could assume that it is a defect free sample; but the pixel amplitude plot reveals damage.

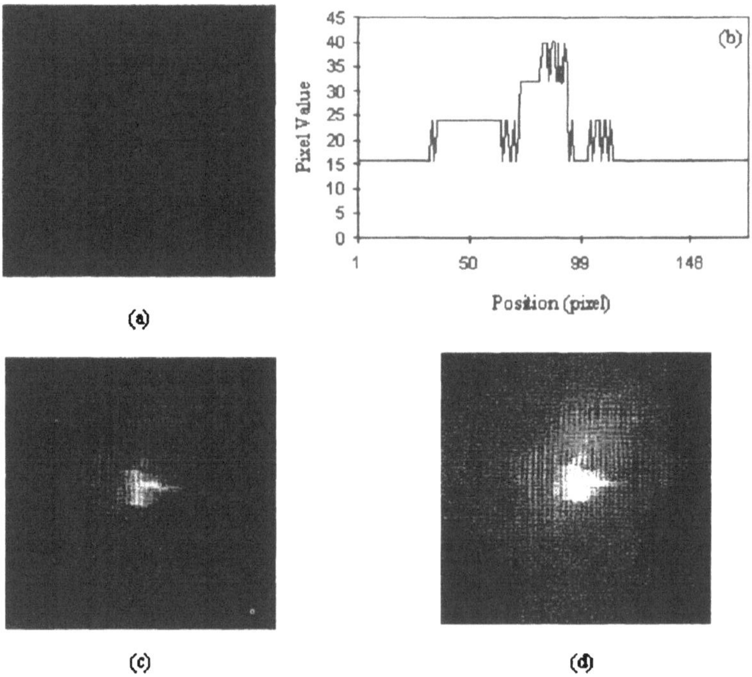

Fig. 3.14 Result of the Bayesian approach on the eddy current and infrared images for increasing values of p(d) and for $p(pv/d)_{ec} = 0.7$ and $p(pv/d)_{ir} = 0.5$; (a): p(d) = 0.1; (b): pixel amplitude plot along the centreline of image (a); (c): p(d) = 0.5; (d): p(d) = 0.9

As p(d) increases from 0.5 (figure 3.14c) to 0.9 (figure 3.14d), the fused image conveys refined information regarding the potential defect area. Moreover, because of the high value of $p(pv/d)_{ec}$, the system behaves in such a manner that it assumes the eddy current image more accurate than the thermograph and thus emphasises this feature on the fused image.

It would be interesting to see how the fused image will evolve if more knowledge is associated with the infrared thermography. Figure 3.15 shows results obtained by increasing the conditional probability associated with the infrared thermographic inspection, assuming that p(d) = 0.5 and $p(pv/d)_{ec} = 0.7$. Figures 3.15a, 3.15b and 3.15c are the results of such a computation for $p(pv/d)_{ir} = 0.2$, 0.5 and 0.7 respectively. Still, there is more knowledge gained towards the eddy current image interpretation but as $p(pv/d)_{ir}$ increases and reaches 0.7, an equal amount of knowledge from both techniques is considered. The fused image conveys less information in favour of the eddy current inspection than was the case in figure 3.14d. The pixel amplitude plots along the centreline of each image also show that the damaged area becomes more distinguishable, despite increasing noise. A similar operation was performed on the two ultrasonic images and is summarised in figure 3.16. In all cases, p(d) = 0.5. The Bayesian operation is performed by automatically assigning p(s/d) based on pixel amplitude value (figure 3.16a). Indeed, the pixel level fusion approach was carried out by considering the grey value of each pixel on images to be fused. It was assumed that a white pixel (i.e. of value 255) has a probability of 1, and a black pixel (i.e. of value 0) a 0 probability. Figure 3.16b was obtained by setting p(pv/d) to 0.3 for the Hitachi ultrasonic image and 0.5 for Krautkramer C-scan. Figure 3.16c results from p(s/d) equal to 0.1 for the Hitachi ultrasonic image and 0.4 for Krautkramer C-scan. In this case, although one may be able to distinguish impact damage more clearly on the final image, and fibres progressively disappear, the pixel amplitude plot still indicates a high noise level. However, if one fuses the eddy current, infrared and Hitachi ultrasonic images, with Bayes's theorem, a S/N ratio improvement is easily visible (figure 3.17). Details on assigning p(pv/d) values for each image of which the fusion results are shown in figure 3.7 are presented in table 3.2.

Table 3.2 Values of p(*pv*/*d*) assigned to the NDT images of which fusion results are shown in figure 3.17

Figure number	Eddy current	Hitachi ultrasonic	Infrared thermography
3.17a	based on grey level	based on grey level	based on grey level
3.17b	0.70	0.30	0.40
3.17c	0.85	0.10	0.25

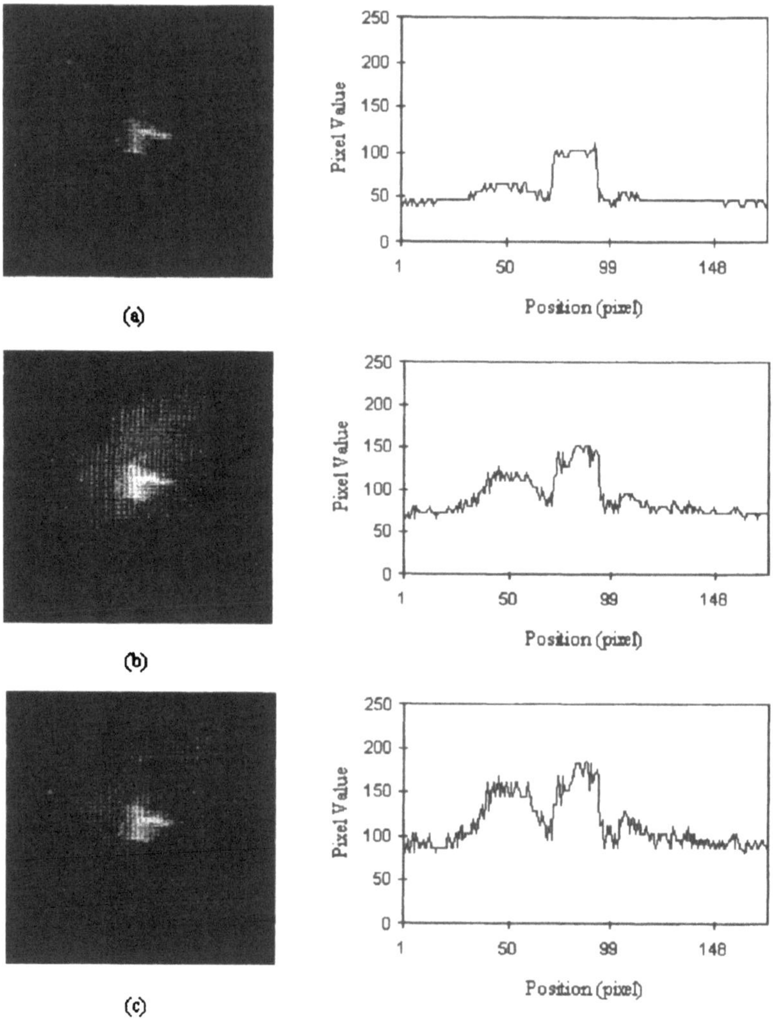

Fig. 3.15 Result of the Bayesian approach on the eddy current and infrared images for $p(d) = 0.5$ and for $p(pv/d)_{ec} = 0.7$;
(a): $p(pv/d)_{ir} = 0.2$; (b): $p(pv/d)_{ir} = 0.5$; (c): $p(pv/d)_{ir} = 0.7$

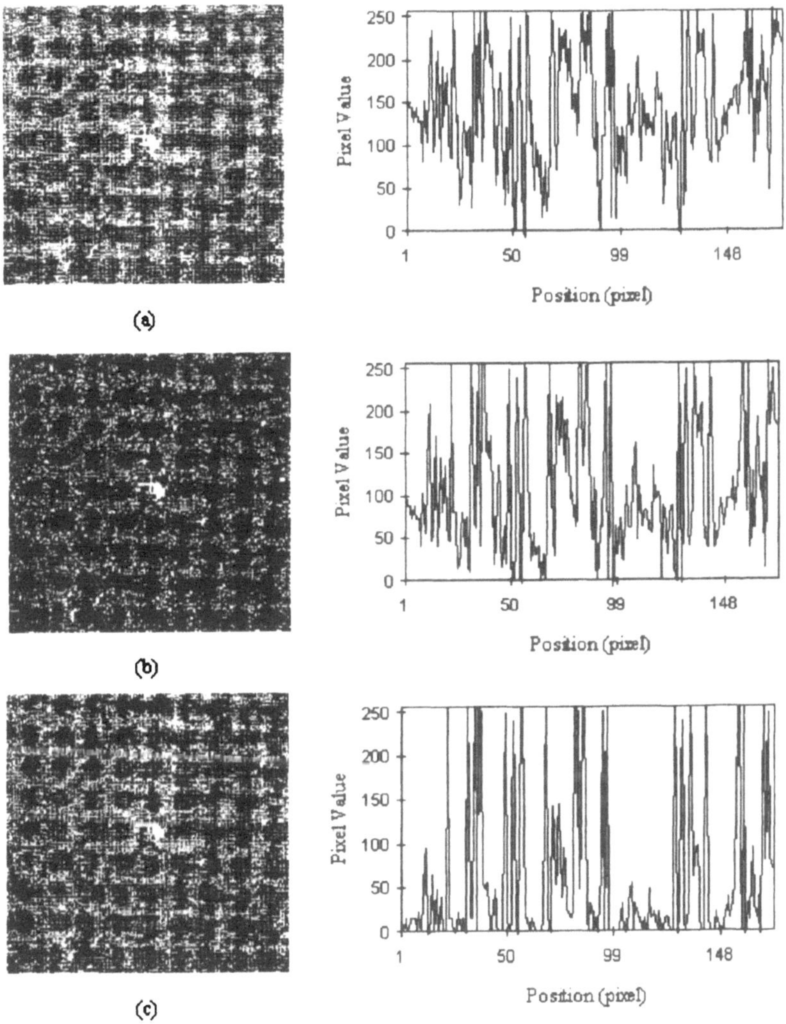

Fig. 3.16 Result of Bayes's theorem on the two ultrasonic images for p(d) = 0.5; (a): p(pv/d) based on grey level; (b): $p(pv/d)_{uth} = 0.3$ and $p(pv/d)_{utk} = 0.5$; (c): $p(pv/d)_{uth} = 0.1$ and $p(pv/d)_{utk} = 0.4$

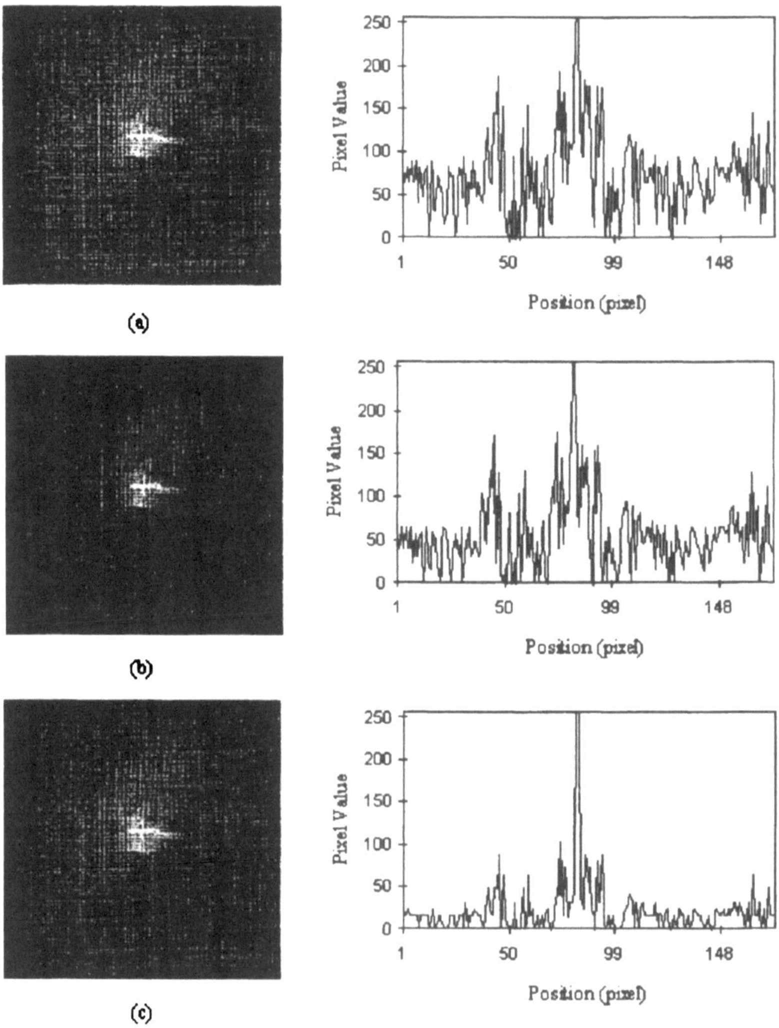

Fig. 3.17 Result of Bayes's theorem on the eddy current, infrared and Hitachi ultrasonic images for p(d) = 0.5 and for p(pv/d) values as described in table 3.2

3.3.6 Dempster-Shafer Theory

Evidential reasoning with the Dempster-Shafer theory can be defined as an extension of the probability theory or a generalisation of Bayes's theorem [15,18]. The Dempster-Shafer theory has already found several applications in NDT data fusion [21]. It has been used in pixel level fusion of X-radiographs, using a grey level variation approach to differentiate between three hypotheses, and make inference on material thickness by selecting the fusion output favoured by the Dempster-Shafer theory [22]. Dupuis *et al.* demonstrated the ability of the Dempster-Shafer approach to help in decision making by combining ultrasonic and radiographic information [23].

In this section, pixel level data fusion using the Dempster-Shafer theory was performed by assigning a degree of belief to a hypothesis and combining information from multiple NDT systems. A mass probability is assigned to each image. These probabilities are integrated to generate a total degree of belief that either supports or refutes the information provided by each NDT method. Such operation is performed through the Dempster's rule of combination; of which the output is an evidential interval that conveys a belief and a plausibility associated with an inspection. A frame of discernment, noted Θ, which is a set of mutually exclusive events, has to be defined. The Dempster's rule of combination [18] is an orthogonal sum which, for eddy current and infrared thermography, can be mathematically expressed as:

$$m_1 \oplus m_2(z) = k \sum_{x \cap y = z} m_{ec}(x) m_{ir}(y) \qquad (3.5)$$

where $m(x)$ is a mass probability, and $m_{ec}(x)$ and $m_{ir}(y)$ the mass probabilities associated with the eddy current and infrared systems respectively. Table 3.3 is an example of the Dempster-Shafer rule of combination of eddy current and infrared information for $m_{ec}(x) = 0.7$ and $m_{ir}(y) = 0.5$. From a similar operation and after normalisation it is possible to establish the evidential interval associated with the resulting images. Figure 3.18 is the result of the pixel level data fusion process using the Dempster-Shafer theory of evidence applied on the eddy current and thermographic images. By using weights of evidence, it is possible to express ignorance and uncertainty as well as belief and plausibility of the fused image. From table 3.3, one can see that there is 35% evidence that the information contained in figure 3.18 is correct. The fused image contains 'filtered' details resulting from the association of weights. In order to identify the damaged area more clearly, we subtracted this image from the original eddy current one. The result of this operation is shown in figure 3.19 revealing the damaged area with the highest confidence level.

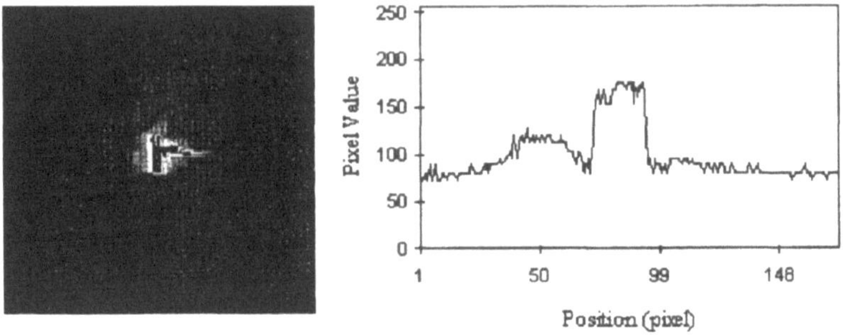

Fig. 3.18 Result of the eddy current and infrared image fusion with the Dempster-Shafer theory of evidence

Table 3.3 Results of the Dempster-Shafer combination of eddy current and infrared thermographic information

	$m_{ec}(x) = 0.7$	$m_{ec}(\Theta) = 0.3$
$m_{ir}(y) = 0.5$	$EC \cap IR = 0.35$	0.15
$m_{ir}(\Theta) = 0.5$	0.35	0.15

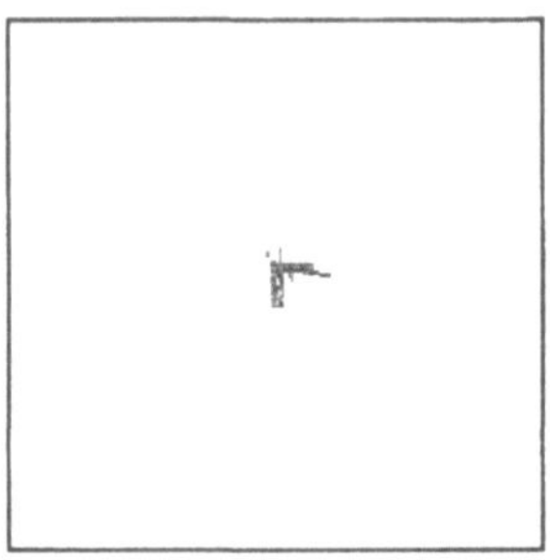

Fig. 3.19 Result of the eddy current and infrared image fusion with the Dempster-Shafer theory of evidence (after image processing of fig. 3.18)

3.3.7 Wavelet and Pyramidal Techniques

The development of the wavelet theory and its popularity has turned it into a fashionable tool for multi-scale decomposition. The wavelet theory is often considered as a special type of pyramid decomposition. Although Fourier transform remains the best approach to determine specific discrete harmonic structure, wavelet approaches that break down time-series signals into time and scale, referred to as 'wavelet space', are now common signal processing operations in science and technology [24].

Wavelet analysis is a powerful signal processing tool to improve the S/N ratio of NDT signals. It has been used to facilitate defect detection and discrimination as well as for image filtering in the reconstruction process of ultrasonic images [25]. Veroy *at al.* used the Morlet wavelet transform to extract and reconstruct Lamb wave dispersion curves from simulated broad-band, multi-mode signals [26].

Because the use of conventional image fusion techniques on raw images seems to reduce contrast and affect the overall quality of an image, pyramid transform approaches were developed in the 80's to carry out fusion in the transform domain by constructing a pyramid transform of a fused image from the pyramid transforms of the source images. The fused image is obtained by taking the inverse pyramid transform. This approach has the advantage of providing images with sharper contrast, as well as to provide both spatial and frequency domain localisation. Different pyramid decomposition algorithms exist such as the Laplacian pyramid, the ratio-of-low-pass pyramid and the gradient pyramid. During the last few years, wavelet-based methods gained increasing popularity and have been widely applied to image processing and analysis [27]. With this technique, the image is represented as a sum of shifted, scaled, dilated kernel functions. The sub-bands are obtained by convoluting and sub-sampling operations. This yields a set of sub-band images of different sizes corresponding to different frequency bands. Probably one of the simplest image fusion methods is to average the two images pixel-by-pixel, but this will lead to a feature contrast reduction. A feature-based selection rule is preferred to perform the fusion operation. It was also demonstrated that it would be better to perform fusion in the transform domain because a pyramid transform can provide information on the sharp construct changes, and can provide both spatial and frequency domain localisation [28]. Because the larger absolute transform values correspond to sharper changes in brightness and thus to salient features in the image such as edges, lines, and region boundaries, the fusion rule can be a selection of the larger absolute value of the two coefficients at the same point. Furthermore, a more complicated fusion rule can be considered, based on the fact that the human visual system detects features at points where the local energy reaches its maximum [29]. The local energy can be defined as the sum of the squared responses of a quadrature pair of odd-symmetric and even-symmetric filters. The most commonly used methods include the pixel-based fusion rule, the window-based fusion rule, and the region-based fusion rule [30]. Correlation of the pixels may occur in their neighbourhood and regions and such information may be used to guide the fusion. The procedure of pixel level multi-resolution image fusion is shown in figure 3.20. A minimum of two images is input prior to the multi-scale/multi-resolution pyramid transform. The sub-images (coefficients) are obtained in the transform domain. The selection varies depending on the applications; Laplacian, ratio-of-low-pass filter, gradient, steerable pyramid or orthonormal wavelet. Each transform has the ability to extract a specific feature of the image. Wavelet can be described as a special type of multi-scale pyramid. In the case of wavelet, the larger absolute value corresponds to sharper changes in brightness and thus to salient features contained in the image such as lines, edges and region boundaries. The fusion of these coefficients is performed in the next stage, using pixel-based and region-based fusion rules. A new set of coefficients results from this operation that can be obtained with an inverse transformation.

Akerman used pyramidal techniques for multisensor fusion applications [31]. He performed a Gaussian pyramid decomposition by low-pass filtering and sampling the result at half the original sample density to obtain a reduced version of the original image. Filtering and decimation is repeated until the size of the image prevents further processing. A Laplacian pyramid is generated by subtracting subsequent levels of the Gaussian pyramid from the level below. The result of the subtraction process used in forming the Laplacian pyramid is to create image planes which contain the difference of the convolutions of two Gaussian weighting functions with the original image. Each Laplacian image plane contains an edge map for features in the corresponding Gaussian image plane. Fusion of Laplacian and Gaussian image planes is then performed at pixel level. The result is a Laplacian pyramid that contains selected information from each sensor. By reversing the image decomposition process used to form the Gaussian and Laplacian pyramids, a single fused image can be obtained.

Multi-resolution analysis with the wavelet transform requires a set of nested multi-resolution sub-spaces. An original space can be decomposed into a lower resolution sub-space, which in turn can be decomposed into another sub-space, and so on (figure 3.21). Analysis of these sub-spaces is easier than the analysis of the original signal itself. A corresponding frequency space can be drawn, and similarly to a space, this frequency band can be divided into sub-bands. In the case of a 2-D image, a 2-D wavelet transform (WT) can be applied. Figure 3.21 shows the structure of a 2-D WT with three decomposition levels.

A schematic of a wavelet image fusion procedure is shown in figure 3.22. A wavelet transform is first performed on each input image prior to the generation of a fusion decision map. The output of the fusion decision map is a fused wavelet coefficient map. An inverse wavelet transform is necessary to obtain the final fused image. In our application we used a pixel-based fusion rule, i.e. the fusion rule applies to the value of each pixel of the whole image (rather than applying a window-based rule to a set of pixels and the fact that there is usually a high correlation among neighbouring pixels). The fusion decision map has the same size as the original image. Each value is the index of the source image which may be more informative on the corresponding wavelet coefficient, and a decision will be made on each coefficient (by considering a pixel value). The pixel-based fusion rule is considered as one of the most straightforward operations, and can be mathematically written as follows:

$$T(C)_L(i,j) = \begin{cases} T(A)_L(i,j), & \text{if } |T(A)_L(i,j)| \geq |T(B)_{L(i,j)}| \\ T(B)_L(i,j), & \text{else} \end{cases} \tag{3.6}$$

where $T(C)_L(i,j)$ is a term that defines the coefficients of the fused image *C* at (i,j) of level *L*. *A* and *B* are the two images to be fused. The coefficients with larger absolute values are involved in the new coefficient set, and finally an inverse transformation is carried out. Similar to a FFT and IFFT, the new set of coefficients is obtained in the transform domain. As shown in figure 3.21 (and figure 3.23 for the Daubechi wavelet), the reconstruction filters (low and high pass filters) are involved in the reconstruction operation. However, because the steerable pyramid is self-inverting, identical filters are used for the reconstructing operation.

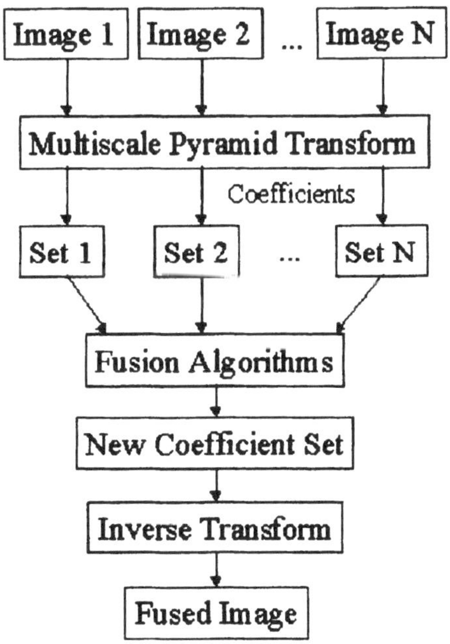

Fig. 3.20 Schematic of the procedure for multi-resolution image fusion

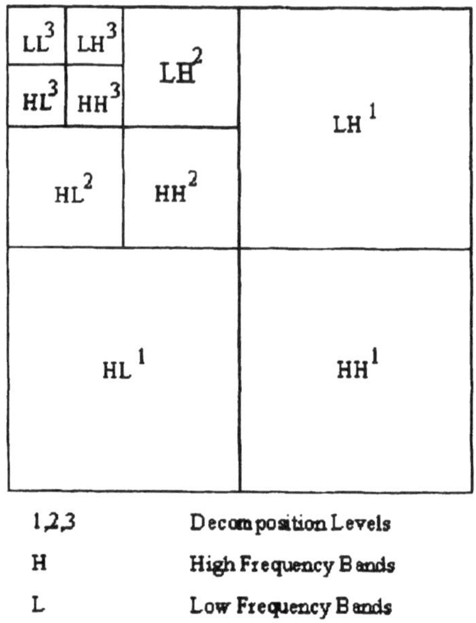

Fig. 3.21 Schematic of 2-D wavelet decomposition levels

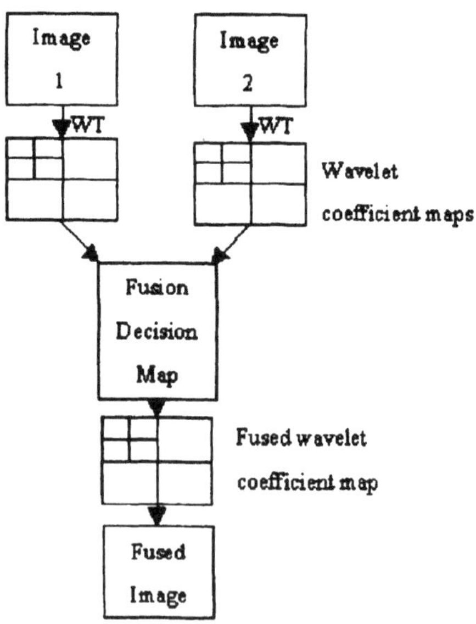

Fig. 3.22 Schematic of a wavelet image fusion procedure

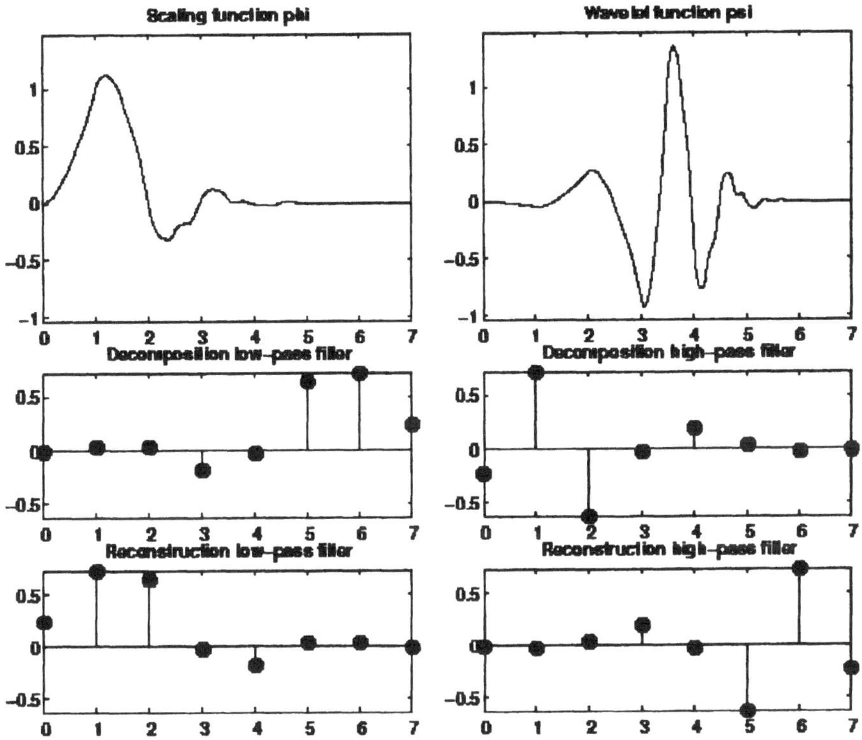

Fig. 3.23 Schematic diagram of the Daubechi wavelet decomposition

Daubechie Wavelet

The Daubechie wavelet 4 (DB4) is a compactly supported orthonormal wavelet function [33]. In figure 3.23, the scaling function $\phi(t)$ and the wavelet function $\varphi(t)$ are given. High-pass $h(t), \tilde{h}(t)$, and low-pass filters $l(t), \tilde{l}(t)$ are also represented. The maximum values of these filters are shown in table 3.4.

Table 3.4 Maximum values of high-pass and low-pass decomposition and reconstruction filters of Daubechie wavelet 4

h	-0.0106	0.0329	0.0308	-0.1870
l	-0.2304	0.7148	-0.6309	-0.0280
h	-0.0280	0.6309	0.7148	0.2304
l	0.1870	0.0308	-0.0329	-0.0106
$\tilde{h}$	0.2304	0.7148	0.6309	-0.0280
$\tilde{l}$	-0.0106	-0.0329	0.0308	0.1870
$\tilde{h}$	-0.1870	0.0308	0.0329	0.0106
$\tilde{l}$	-0.0280	-0.6309	0.7148	-0.2304

The relation between these filters, the scaling function and the wavelet function are as follows:

$$\phi(t) = \sum_n h_n \phi(2t - n) \tag{3.7}$$

$$\tilde{\phi}(t) = \sum_n \tilde{h}_n \tilde{\phi}(2t - n) \tag{3.8}$$

$$\varphi(t) = \sum_n l_n \varphi(2t - n) \tag{3.9}$$

$$\tilde{\varphi}(t) = \sum_n \tilde{l}_n \tilde{\varphi}(2t - n) \tag{3.10}$$

The relation between the filters involved can be estimated from table 3.4:

$$\tilde{h}(n) = h(-n) \tag{3.11}$$

$$\tilde{l}(n) = l(-n) \tag{3.12}$$

$$l(n) = (-1)^{1-n} h(1-n) \tag{3.13}$$

The DB4 theory was used to fuse the eddy current and infrared images previously described. Two images are decomposed and reconstructed as shown in figure 3.22. Figure 3.24 shows the fusion result using the DB4 wavelet theory. One notices a reduction in noise compared with the original images. It was found that the texture of the surface of the material seems to affect the fusion operation as seen on the contour mapping, and makes the extent of impact damage difficult to define.

Fig. 3.24 Fused image obtained with the Daubechi wavelet and its contour plot

Steerable pyramid transform

The first level of the steerable pyramid decomposition is given in figure 3.25. On this figure, $H0$ is a high-pass filter, $L0$ and $L1$ are low-pass filters while $B0$-$B3$ are band-pass filters. The steerable pyramid can be described as a multi-scale, multi-orientation image decomposition approach. It is known that edge and line information can be obtained by derivative-based operations. the steerable pyramid is such a computation, which combines the multi-scale decomposition with differential measurements. It can capture some of the oriented structure in images. It is also non-orthogonal and over-complete. Additionally, the steerable pyramid is self-inverting, that is, the decomposition and reconstruction filters are the same. A Matlab code implementation of the steerable pyramid is available on the World Wide Web at http://www.cis.upenn.edu/~eero/home.html, and references can provide more information about the steerable pyramid transform [34-35]. Figure 3.26 shows the result of the fusion operation using the steerable pyramid technique. Based on the contour plot, it appears that this fused image contains more complete information about the extent of defect damage. It is also found that the texture of the material affects the fusion output.

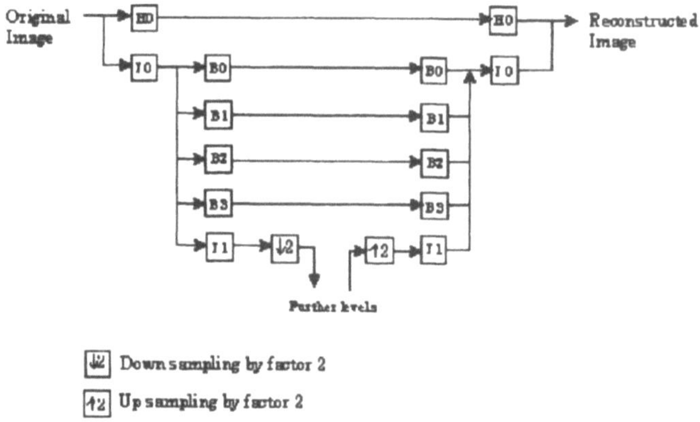

Fig. 3.25 First level of the steerable pyramid decomposition

Fig. 3.26 Result of the fusion operation using the steerable pyramid technique on the eddy current and infrared thermographic images, and contour plot

Application of multi-resolution mosaic

Image segmentation is usually adopted to extract the objects of interest from background in a given image. In the case of the eddy current image, it is relatively easy to perform a segmentation operation. However, the texture of the surface of the material as detected with infrared thermography will affect the threshold operation, and a pre-processing operation is needed to reduce noise. First, a de-noising operation is applied which is followed by a 5×5 Gaussian filtering operation. Then threshold and edge contouring operations are carried out. These operations are implemented using Matlab Wavelet Toolbox and ImageTool developed by the University of Texas Health Science Center in San Antonio, USA. The threshold operation is usually used to separate the object from its background, in such a way that the regions of interest from each image can be added [36].

Mosaic techniques have been used to combine two or more signals into a new one with an invisible smoothing seam [37]. The seam between the two parts will be smoothed and the characteristics of each one is still preserved (it makes the edges between the two parts indistinguishable). Thus, the related parts in the images may be extracted and combined using the mosaic technique. The second step is to segment the new image. Four steps in wavelet based mosaic technique were adopted and are presented below [37]:

- Step 1: Perform 2-D wavelet decomposition of two images *A* and *B*;
- Step 2: Chose a low-pass filter $h = \{0.25, 0.5, 0.25\}$, and use this filter to generate a sequence of low-passed and sub-sampled signals of the mask signal *S* (figure 3.27a): $W_{2^k}(S)$ $(1 \le k \le L)$;
- Step 3: Construct a sequence of signals using $W_{2^k}(S)$ as weighting functions in the corresponding resolution level (as shown in equation 3.14) [38];
- Step 4: Perform an inverse wavelet transformation to obtain the mosaic signal *C* (figure 3.27b).

$$\begin{aligned} A^d_{2^L} f(C) &= A^d_{2^L} f(A) \cdot W_{2^L}(S) + A^d_{2^L} f(B) \cdot (1 - W_{2^L}(S)) \\ D^j_{2^k} f(C) &= D^j_{2^k} f(A) \cdot W_{2^k}(S) + D^j_{2^k} f(B) \cdot (1 - W_{2^k}(S)) \\ & \qquad 1 \le k \le L \quad 1 \le j \le 3 \end{aligned} \tag{3.14}$$

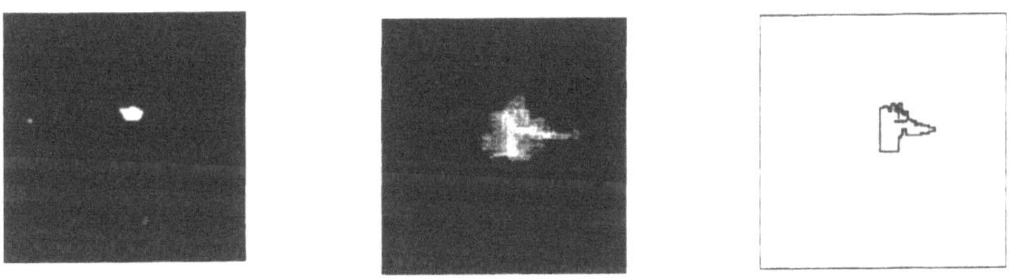

(a) (b) (c)

Fig. 3.27 (a): Mask image; (b): mosaic result; (c): contour image of (b)

3.4 DISCUSSION

The data fusion results at pixel level of a synthetic image from an eddy current inspection, ultrasonic C-scans, and a real image from an infrared thermographic examination have been considered to determine some of the advantages and limitations of various data fusion procedures. The results of the data fusion processes have been assessed qualitatively and quantitatively [39]. The quality of the fused image was estimated in function of its sharpness for ease of interpretation and defect location. In order to quantify the results of the data fusion process, the area of impact damage was calculated for each fused image [39]. However, such results are of no interest if one does not have a reference value to compare them with. Such values can only be obtained either by fractography (this may be difficult to obtain due to the mechanical structure of the composite material) or by assuming that the inspection result from one NDT technique provides the most likely representation of the actual damaged area. Even so, such approximation and comparison should be regarded as a simple estimation procedure. Nevertheless, the previous experiments showed that pixel level data fusion can enhance defect detection. Indeed, image processing operations such as edge detection and tracing of contours can increase defect location provided the original image has good resolution and sharpness. These operations find better applications in the field of pattern recognition for automated defect location. Maximum amplitude selection can be very useful for defect location provided that information from each system is displayed in an identical image format. Adding both images using an AND operator can be an easy means to improve the S/N ratio. Ensemble averaging is better suited to reduce noise in multiple images from the same source. Simple averaging of the four images did not provide a gain or loss in qualitative information. Weighted averaging may be more adequate than ensemble averaging, but the result of the operation depends upon the weight associated with each technique. Nevertheless, it is adequate to reduce spurious information which may result from the use of a technique with limited efficiency [20]. The weights need to be estimated from previous knowledge about the system used and the material tested to avoid the cancellation of useful details. Previous NDE and POD curves constitute the basis for choosing weight factors for weighted averaging and mass probabilities for the Dempster-Shafer theory [15]. These POD curves were plotted for multiple NDT techniques (e.g. visual, eddy current, radiography and infrared thermography) used for the detection and quantification of impact damage in composites (a comprehensive description of this operation can be found in the literature [15]).

The Bayesian approach is best suited to test for hypotheses and to combine information from one, two or more sources. The output of the Bayesian approach is affected by the selection of the prior probability which should be derived from previous experiments. Associating a prior probability based on grey level values is arbitrary and could lead to errors. It was also observed that computational time may increase rapidly if a large number of sources of information are considered [15]. One of the advantages of the Bayesian approach is that ambiguity can be removed and more knowledge added towards the NDT system considered the most efficient. But, this may result in very little use of the information from a system with lower performances (in terms of defect detection and characterisation), sometimes providing misleading outputs [15].

A difference between the Dempster-Shafer theory and the Bayesian theory is that if there is insufficient knowledge available, the Dempster-Shafer theory does not assign belief or ignorance to the fused image. The output of the

Dempster-Shafer rule of combination is also greatly dependent upon the value of the mass probabilities and small changes may give incorrect information. Such probabilities would have to be carefully thought about prior to the data fusion operation. If the mass assigned to the image is high, the process tends to neglect information from the other NDT method and assigns a degree of belief towards the resulting image. The use of segmentation with a threshold value may provide a more accurate dimensioning of the damage. The steerable pyramid offers an alternative to conventional fusion approaches. The multi-resolution mosaic approach appears as a relatively robust method to provide a consensus assessment of the actual damaged area. Also, a fixed threshold operation may not fit the entire image, an adaptive thresholding would have to be considered for pre-processing [40].

The data fusion techniques investigated for the fusion of images from different NDT systems - despite being adequate for the fusion of other types of images (as reported in the literature) - need additional refinement. This is caused by the disparity in image quality that may result between various NDT systems. Indeed, in the case of the eddy current inspection, the resulting defect image was easily recognisable and the woven structure of the material did not interfere with the output. On the other hand, the infrared thermograph did not clearly reveal the position of the impact damage, the output image depended greatly on the external condition (e.g. external source of heat and parasite light), and the woven structure of the material interfered with the analysis of the image. Such parameters should be considered for each type of NDT image prior to fusion. For example, a pre-processing operation should be performed to remove the woven structure as detected with the infrared thermographic camera. Ultrasonic images revealed additional artefacts and would therefore require an additional form of pre-processing prior to fusion, and very probably the use of a different mathematical algorithm for a fusion operation more adapted to the combination of different ultrasonic images.

3.5 CONCLUSION

The results obtained from pixel level data fusion operations performed using eddy current, ultrasonic and infrared thermographic images demonstrated the potential and some of the limitations of a few mathematical algorithms. Although additional knowledge about the 3-D structure of a material would be desirable for materials evaluation, further research is necessary to fully determine the most adequate fusion strategy. Still the data fusion process depends greatly upon the problem to be considered, and current experiments for the fusion of information from several sources suggest that satisfactory data fusion processes for the combination of two images may not be adequate when the set of data is modified. Because of this limitation, a data fusion architecture would have to be designed.

This may be costly or time consuming. Nevertheless, on a long term basis it is thought that such development will benefit the whole NDT community as it will help improve safety standards through a more careful and accurate characterisation of flaws and defects. It is believed that more advanced pixel level information fusion techniques may be developed for specific non-destructive applications at the manufacturing level.

REFERENCES

1. Black J.V., *Fusion of infrared and visible-light images*, Proc. 1st Command Info. Syst. Workshop, 1994, USA, 316-325.
2. Reed J.M., Hutchinson S., *Image fusion and subpixel parameter estimation for automated optical inspection of electronic component*, IEEE Trans. on Industrial Electronics, 1996, 43(3):346-354.
3. Dromigny A., Zhu Y.M., *Improving the dynamic range of real-time x-ray imaging systems via Bayesian fusion*, J. of Nondestructive Evaluation, 1997, 16(3):147-159.
4. Mina M., Udpa S.S., Udpa L., Yim J., *A new approach for practical two dimensional data fusion utilizing a single eddy current probe*, Proc. Review of Progress in QNDE, 1997, Brunswick, USA, 16A:749-755.
5. Sun K., Udpa S., Udpa L., Xue T., Lord W., *Registration issues in the fusion of eddy current and ultrasound NDE data using Q-transforms*, Proc. Review of Progress in QNDE, 1996, Seattle, USA, 15A:813-820.
6. Barniv Y., Casasent D., *Multisensor image registration: experimental verification*, Proc. SPIE, Processing of Images and Data from Optical Sensors, 1981, USA, 292:160-171.
7. Jain A.K., Dubuisson M.P., Madhukar M.S., *Multi-sensor fusion for nondestructive inspection of fiber reinforced composite materials*, Proc. 6th Tech. Conf. of the American Society for Composites, 1991, Albany, USA, 941-950.
8. Yanowitz S.D., Bruckstein A.M., *A new method for image segmentation*, Computer Vision, Graphics and Image Processing, 1989, 46(4):82-95.
9. Scarponi C., Briotti G., *Ultrasonic detection of delamination on composite materials*, J. of Reinforced Plastics and Composites, 1997, 16(9):768-790.
10. Broughton W.R., Sims G.D., M.J. Lodeiro M.J., *Overview of DTI-funded programme on 'Standardised procedures for ultrasonic inspection of polymer matrix composites'*, Insight, 1998, 40(1):8-11.
11. Gros X.E., Takahashi K., *A comparison of impact damage on thermoplastic toughened and conventional CFRP materials by ultrasonic examination*, Proc. of the Materials & Mechanics Committee of Japan Society of Mechanical Engineers, 1998, Kumamoto, Japan, 413-414.
12. Gros X.E., *Characterization of low energy impact damages in composites*, J. of Reinforced Plastics and Composites, 1996, 15(3):267-282.
13. Gros X.E., *Eddy current testing: the future of composite materials evaluation ?*, J. of JSNDI, 1997, 46(9):642-648.
14. Lowden D.W., Gros X.E., Strachan P., *Visualising Defect Geometry in Composite Materials*, Proc. Inter. Symp. on Advanced Materials for Lightweight Structures, 1994, Netherlands, 683-686.
15. Gros X.E., *NDT Data Fusion*, Butterworth-Heinemann, London, 1997.
16. Abbate A., Frankel J., Das P., *Wavelet transform signal processing applied to ultrasonics*, Proc. Review of Progress in QNDE, 1996, Seattle, USA, 15A:741-748.
17. Hays W.L., Winkler R.L., *Statistics, Probability, Inference, and Decision*, II, HRW Pub., 1970.
18. Shafer G., *A mathematical theory of evidence*, Princeton Uni. Press, USA, 1976.
19. Zadeh L.A., *The role of fuzzy logic in the management of uncertainty in expert systems*, Fuzzy Sets and Systems, 1983, 11:199-227.
20. Gros X.E., *A review of NDT data fusion processes and applications*, Insight, 1997, 39(9):652-657.
21. Dromigny-Badin A., Rossato S., Zhu Y.M., *Fusion de données radioscopiques et ultrasonores via la théorie de l'évidence*, Traitement du Signal, 1997, 14(5):499-510.

22. Dromigny-Badin A., Zhu Y.M., *Fusion de données complémentaires en vue de l'amélioration de la dynamique des systèmes d'imagerie*, Proc. 16th GRETSI Conf., 1997, Grenoble, France, 1415-1418.
23. Dupuis O., Kaftandjian V., Zhu Y.M., Babot D., *Détection de défauts par fusion de signaux ultrasonores et d'images radiographiques*, Proc. 17th GRETSI Conf., 1999, Vannes, France.
24. Graps A., *Wavelets are stars in astronomy problems*, Scientific Computing World, 1998, (37):22-23.
25. Lasaygues P., Lefebvre J.P., *Application de l'analyse en ondelettes en contrôle non destructif par ultrasons*, Proc. 6th Europ. Conf. on NDT, 1994, Nice, France, 1:19-23.
26. Veroy K.L., Wooh S.C., Shi Y., *Analysis of dispersive waves using the wavelet transform*, Proc. Review of Progress in QNDE, 1999, Snowbird, USA, 18A:687-694.
27. Mallat S.G., *A theory for multiresolution signal decomposition: the wavelet representation*, IEEE Trans. on Pattern Analysis and Machine Intelligence, 1989, 11(7):674-693.
28. Li H., Manjunath B.S., Mitra S.K., *Multisensor image fusion using the wavelet transform*, Graphical Models and Image Processing, 1995, 57(3):235-245.
29. Koren I., Laine A., Taylor F., *Image fusion using steerable dynamic wavelet transform*, Proc. IEEE Inter. Conf. on Image Processing, 1995, Washington, USA, 232-235.
30. Zhang Z., Blum R.S., *Image fusion for a digital camera application*, Proc. 32nd Asilomar Conference on Signals, Systems, and Computers, 1998, Monterey, USA.
31. Akerman A., *Pyramidal techniques for multisensor fusion*, Proc. SPIE Sensor Fusion V, 1992, Boston, USA, 1828:124-131.
32. Nievergelt Y., *Wavelets made easy*, Birkhauser, 1998.
33. Temme N., *An introduction to wavelets*, Wavelet Course, The Netherlands National Research Institute for Mathematics and Computer Science, 1997.
34. Freeman W.T., Adelson E.H., *The design and use of steerable filters*, IEEE Trans. on Pattern Analysis and Machine Intelligence, 1991, 13(9):891-907.
35. Simoncelli E.P., Freeman W.T., Adelson E.H., Heeger D. J., *Shiftable multi-scale transforms*, IEEE Trans. on Information Theory, 1992, 38(2):587-607.
36. Liu Z., Gros X.E., Tsukada K., Hanasaki K., Takahashi K., *The use of wavelets for pixel level NDT data fusion*, Proc. 2nd Japan-US Symp. on Advances in NDT, 1999, Hawaii, USA, 474-477.
37. Hsu C.T., Wu J. L., *Multiresolution mosaic*, IEEE Trans. on Consumer Electronics, 1996, 42(4):981-990.
38. Burt P.J., Adelson E.H., *The Laplacian pyramid as a compact image code*, IEEE Trans. on Communications, 1989, 31(4):532-540.
39. Gros X.E., Bousigué J., Takahashi K., *NDT data fusion at pixel level*, NDT&E International, 1999, 32(5):283-292.
40. Jain A.K., Dubuisson M.P., *Segmentation of X-ray and C-scan images of fiber reinforced composite materials*, Pattern Recognition, 1992, 25(3):257-270.

4 FUZZY LOGIC DATA FUSION FOR CONDITION MONITORING

Pan FU[1], Anthony HOPE[2]
[1]University of Hertfordshire
Hatfield, England
[2]Southampton Institute,
Southampton, England

4.1 INTRODUCTION

Metal cutting operations constitute a large percentage of manufacturing activity. One of the most important objectives of metal cutting research is to develop techniques that enable optimal utilisation of machine tools, improved production efficiency, high machining accuracy, reduced machine downtime and tooling costs. To realise this objective, it is necessary to develop integrated, self-adjusting manufacturing systems that are capable of machining various parts automatically without operator supervision. Cutting tool condition monitoring is certainly an important monitoring requirement of unattended machining operations. Using traditional tool change strategies, tools are either replaced after a given number of shifts or the operator would change the tool when he thought it to be no longer capable of performing normally. Generally speaking tools are under-utilised because most estimated tool life data are taken conservatively to ensure machining quality. However, it is also possible for a worn tool to be continuously used in the machining process so the parts produced do not meet the required accuracy standard. It is therefore necessary that an intelligent sensing system be devised to monitor the tool wear state during cutting operations so that worn tools can be replaced at the optimum time [1-2].

It has now been widely accepted that, under varying machining conditions, the information required to make reliable decisions on the tool wear state is unlikely to be available using single sensor information. Multisensor data fusion is attractive since the loss of sensitivity of one sensor can be compensated by other sensors. In this work, different types of transducers were initially investigated and as a result, power consumption, cutting force, vibration and acoustic emission sensors were chosen as the most appropriate for tool wear monitoring. Features relevant to tool wear state may be drawn from both the time and frequency domains. In this study, time domain features such as mean value, standard deviation, histogram components and kurtosis have been found to be simple but reliable indications of tool wear. Spectral analysis was found to be more promising for cutting force signals, where the energy of a force signal varied within particular frequency ranges as the wear on the cutting tool increased.

It is difficult for a computer-based tool condition monitoring system to simulate the process of the human pattern recognition process, which is highly complex and poorly understood. The monitoring system should be able to process the incoming tool wear signal features and accomplish the pattern recognition process, in which it correlates the incoming features with a particular state of tool wear. In order to obtain a more robust and reliable decision, the pattern recognition process should be carried out in parallel by making use of all the features from different sensor signals simultaneously. Another necessary precondition for pattern recognition is the machine learning algorithm. Such algorithms regulate their learning parameters by studying the signal features corresponding to different tool wear states. The pattern recognition approach provides a framework for machine learning and integration of multisensor information in the manufacturing environment. The introduction of artificial intelligence greatly improves the performance of modern tool condition monitoring systems [3-4]. It has a much greater functionality than a conventional monitoring system since it can respond properly to the different characteristics of the machine tool or process it is monitoring. The following functions are necessary: signal processing, multisensor data fusion, learning and decision making. The basic structure of a modern tool condition monitoring system is illustrated in figure 4.1.

A unique neuro-fuzzy hybrid pattern recognition algorithm was developed to carry out the fusion of multisensor information and signal feature pattern recognition [5]. The neuro-fuzzy system combines the transparent representation of a fuzzy system with the learning ability of neural networks. The algorithm has strong modelling and noise suppression ability. A large number of experiments have been carried out to prove that the proposed data processing algorithm is accurate and reliable under different machining conditions. Modern tool wear monitoring systems are based on advanced sensing techniques and signal processing methods. The sensors pick up tool wear relevant signals (e.g. cutting force, vibration and acoustic emission) from the metal cutting process.

Advanced signal processing techniques and artificial intelligence have been a necessary part of tool wear monitoring systems. Artificial neural networks, fuzzy logic, expert systems and genetic algorithms have all been employed to determine the complex relationships between tool wear states and sensor signals [6-8].

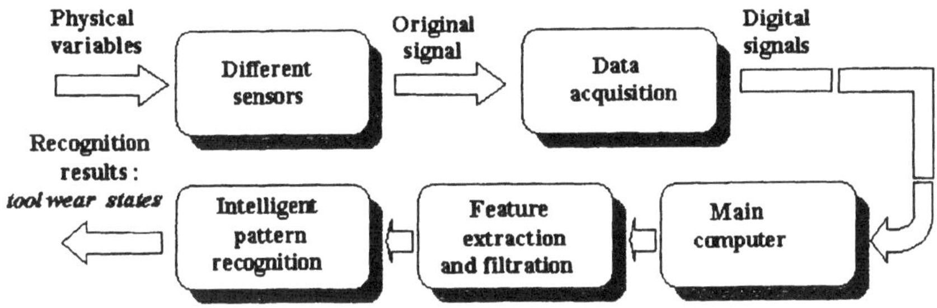

Fig. 4.1 Schematic of a tool condition monitoring system

4.2 PROGRESSIVE TOOL WEAR

A cutting tool wears along the cutting edge and on adjacent surfaces. Figure 4.2 shows how a new tool may wear during the cutting process.

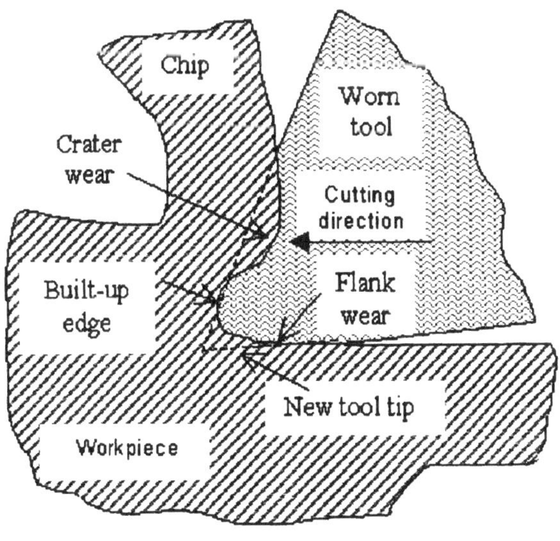

Fig. 4.2 Tool wear surfaces

Progressive wear of a tool may appear in two forms, crater wear and flank wear [9]. In metal cutting, the temperature on the tool face can be as high as 1000°C or more. The tool material may soften rapidly under this high temperature and along the rake face, the chip motion and high normal stress can produce a wear scar called crater wear. In the circumstances when the cutting speed is very high, crater wear is usually the key factor to determine the tool life. Eventually the crater wear will become very severe, hence the tool edge is weakened too much, and the tool fractures. However, when the tool is used under normal economical conditions the effect of crater wear is less significant and flank wear is far more important. Along the clearance surface of the tool, tool motion and high normal stress cause severe friction between workpiece surface and the tool flank, producing flank wear. The width of the wear land is used to measure the amount of wear and can be easily measured by using a microscope. The width of the flank wear land is designated VB and is the distance from the original cutting edge to the limit of the wear area that intersects the flank of the insert. Figure 4.3 shows a typical curve of the progress of flank wear land width, VB *vs.* cutting time (VB is the width of the flank wear land measured in millimetres). The tool wear process can be divided into three sections:

- *Initial breakdown*: the sharp tool edge is quickly broken down and an initial wear land is created.
- *Steady state wear*: the wear land increases at an even rate during this stage.
- *Accelerated wear*: the wear rate rises abruptly when the temperature at the wear land becomes very high.

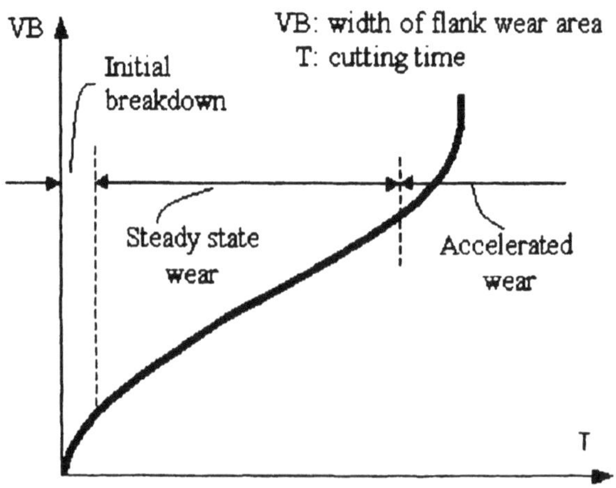

Fig. 4.3 Development of flank wear with time

4.3 INTELLIGENT TOOL WEAR MONITORING TECHNIQUES

A practical on-line tool condition monitoring system should be highly reliable, be able to detect tool wear in a wide range of machining conditions and be sensitive to changes of tool wear states. This leads to the development of intelligent tool wear monitoring approaches. Most modern tool condition sensing systems could be classified as multisensor and intelligent sensing systems. They have a much greater functionality than conventional systems to meet the special needs of the machine tool and process they are monitoring. They are able to utilise experience gained from past operations and accumulate knowledge through learning, so ambiguous information can be accommodated [10-11].

An intelligent sensing system can be easily distinguished from a conventional system as the former can make decisions automatically and the latter must follow predetermined orders. Advanced artificial intelligence-based techniques are the core of these systems. Neural networks, fuzzy logic or genetic algorithms are popular tools for sensor signal characterisation and decision making.

4.3.1 Fuzzy Clustering Algorithms

Clustering analysis is the art of finding groups within data [12-14]. Hard clustering methods form some hard clusters and an object must belong to only one cluster. Fuzzy clustering methods are capable of describing ambiguity in the data and hence are suitable for more realistically identifying the difference of tool wear states. There are two popular fuzzy clustering methods: one is based on the fuzzy relationship between patterns and the other is the fuzzy C-means (FCM) algorithm. In the latter an objective function approach is used for clustering the data into hyper-spherical clusters and the FCM algorithm is realised by iteration, based on an optimal principle. The FCM algorithm is creatively applied in this study to develop a signal feature filter.

4.4 THE CONDITION MONITORING SYSTEM

This section introduces the establishment of a tool wear monitoring system for milling operations. The system includes four kinds of transducers (i.e. power consumption, cutting force, vibration and acoustic emission), signal conditioning devices and a computer. Milling experiments were carried out on a vertical machining centre using three different sets of cutting parameters, namely cutting speed, feed rate and depth of cut.

4.4.1 Measurement of the Power Consumption of the Spindle Motor

In the electrical cabinet of the machining centre, the control unit provided a voltage signal that was proportional to the load rating of the spindle motor. This signal was conveniently used to relate cutting tool condition to the power consumption of the machine tool.

4.4.2 Measurement of Cutting Force

A three component quartz dynamometer was chosen to measure the cutting forces in three mutually perpendicular directions (i.e. x, y and z). This dynamometer can detect very small dynamic changes in large forces. The output signal of the dynamometer was fed through an armoured connecting cable into a three channel charge amplifier and the amplified signals were then transmitted to the signal sampling device.

4.4.3 Vibrations Measurement

It was necessary to mount the vibration sensor as close to the cutting tool as possible to obtain healthy vibration signals. A small, light accelerometer was eventually chosen as it could be mounted directly on the workpiece and combined with a charge amplifier, the vibration sensing system had a wide frequency response from 60 Hz to 30 kHz.

4.4.4 Measurement of Acoustic Emission

The metal cutting process generates elastic stress waves known as acoustic emission (AE) which propagate through the machine structure and are related to the rapid release of strain energy from the deforming material areas. Flank wear is primarily the result of the rubbing action between the tool and the newly formed workpiece and the extent of the wear depends strongly on the rate of metal deformation. The AE measuring apparatus was composed of three parts; an AE sensor with a frequency response from 100 kHz to 600 kHz, a pre-amplifier and an AE output instrument which included an interface module to connect the instrument to an intelligent terminal.

4.4.5 Tool Wear Monitoring System

The tool wear monitoring system consists of four sensors (power consumption, cutting force, vibration and AE), data acquisition devices and a computer, as shown schematically in figure 4.4.

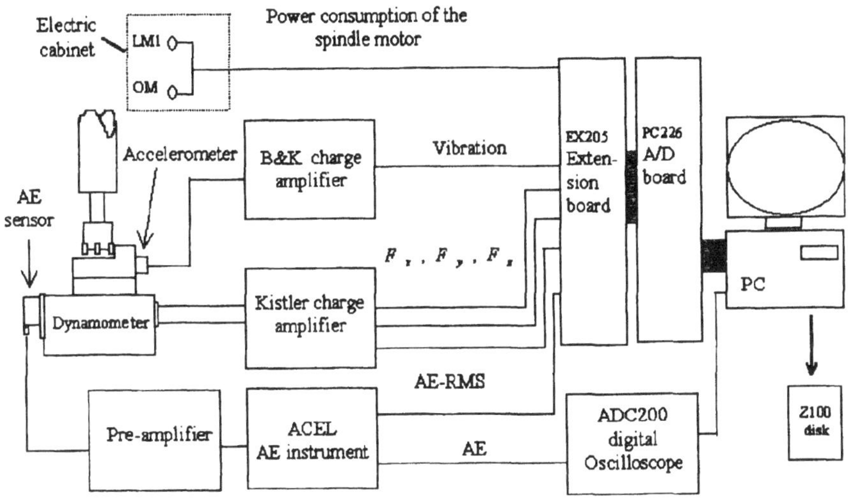

Fig. 4.4 Schematic of the tool wear monitoring system for milling operation

The dynamometer was installed on the table of the machine tool with the workpiece mounted on top of it. Three orthogonal cutting forces were accurately detected and the power consumption signal was obtained directly from the electrical cabinet of the machine tool. Detail of the monitoring system is shown in figure 4.5.

The accelerometer was mounted on the side of the workpiece to monitor the vibrations with minimum loss at the nearest site from the vibration source. The installation of the accelerometer can be seen in figure 4.5. The AE sensor was installed on the side of the dynamometer. To improve transmission a very thin layer of silicon grease was positioned between the AE sensor surface and the side of the dynamometer. Experiments showed that the AE sensor could provide healthy signals in this position. The high frequency AE signal was sampled and placed in the buffer of the ADC200 digital oscilloscope. The digital results were transmitted to the computer and stored in its hard disk. The other six types of signal, i.e. power consumption, vibration, AE-RMS and three cutting force signals were collected by the PC226 A/D board and then stored in the hard disk.

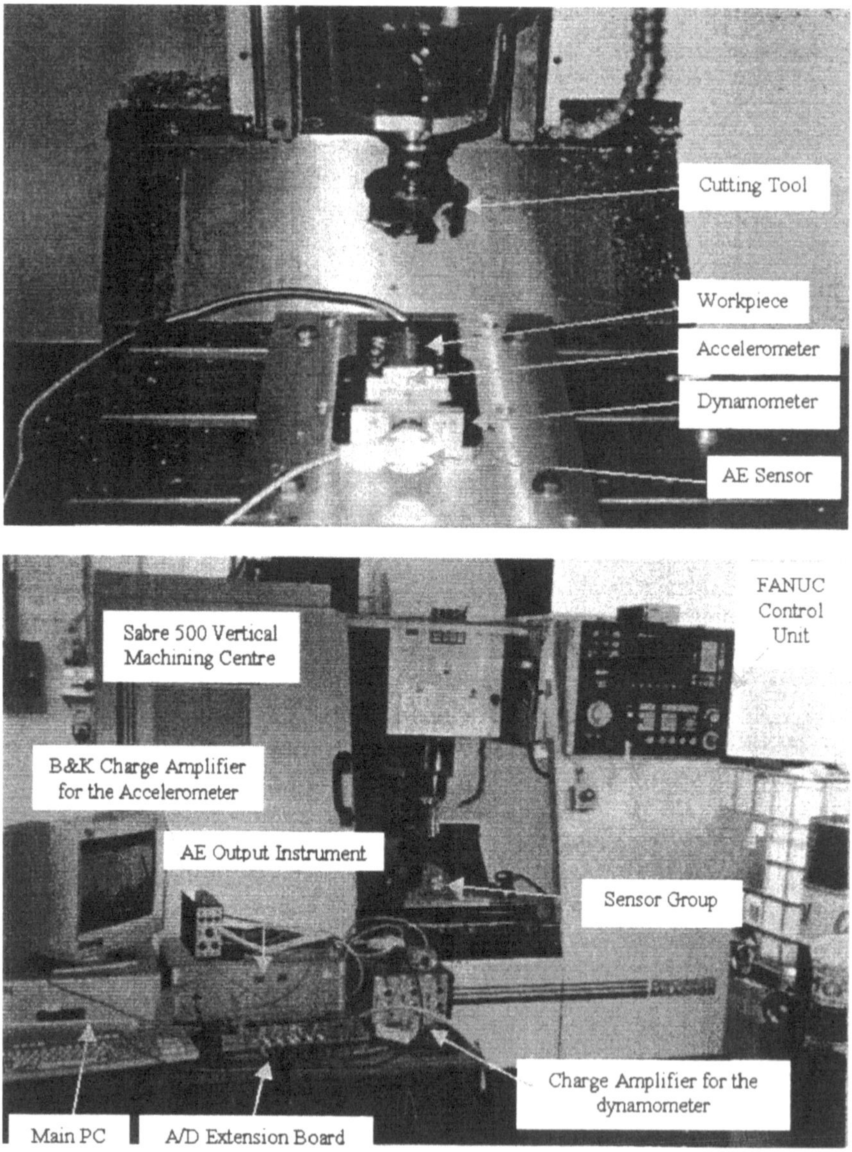

Fig. 4.5 Experimental tool condition monitoring system; the multisensor arrangement (top), on-line multi-channel signal collection system (bottom)

4.5 FEATURES EXTRACTION

The tool wear monitoring system samples power consumption, force, AE and vibration signals from the milling process. The original signals have large dimensions and cannot be directly used to estimate tool wear value. This is because the number of parameters in the neuro-fuzzy networks, analysed to recognise tool wear state, is an exponential function of the input space dimension. Therefore, the cost of implementing the modelling and obtaining the output increases exponentially as the input space dimension grows.

Several useful features can be extracted from the time domain, such as the mean value, which is a simple but effective parameter. For force, vibration and AE-RMS signals, standard deviation, histogram component and kurtosis were also found to change as the width of the flank wear land increased. Spectral analysis in the frequency domain can provide further information, and experimental results showed that for a cutting force signal, the frequency component distribution changed with tool wear development. The spectral analysis results were compressed by dividing the whole spectrum into ten frequency sections and using the average values in each section as the frequency components. Altogether fifty features were extracted from the time domain and frequency domain for further pattern recognition as summarised in table 4.1.

Table 4.1 Features extracted from time and frequency domains

Transducer	Features Extracted
Power consumption	Mean value
Dynamometer	Mean value, standard deviation, 10 frequency components for F_x, F_y, and F_z
Acoustic emission	Mean value and 10 histogram components
Accelerometer	Standard deviation and kurtosis

In this study, all features were normalised to a unified scale (0~1), so that the neuro-fuzzy networks could integrate multisensor information to recognise tool wear state. In the normalising process, for each signal feature (under a specific set of cutting conditions), the biggest sampling value was converted to 0.9 and all the other sampling values were normalised correspondingly. The technique used to extract features in the frequency domain for the cutting force signals is illustrated in the following section.

4.5.1 Cutting Force Signals

The dynamometer sampled three orthogonal cutting forces F_x, F_y, F_z and from the milling operation. These signals varied as the tool flank wear value increased and figure 4.6 shows two typical groups of force signals corresponding to new and worn tools.

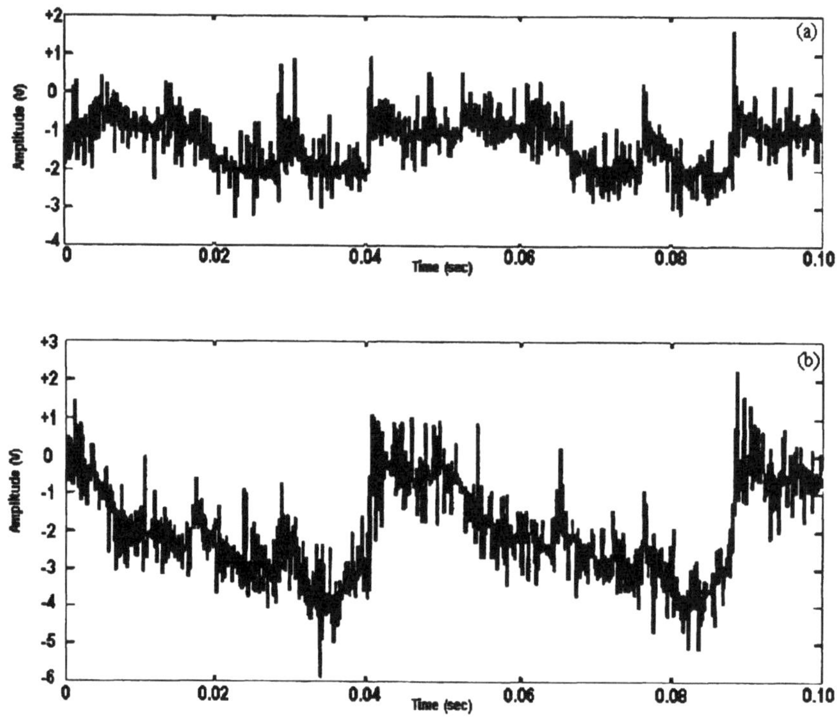

Fig. 4.6 (a) Plot of cutting force Fx *vs*. time for a new tool, and (b) for a worn tool

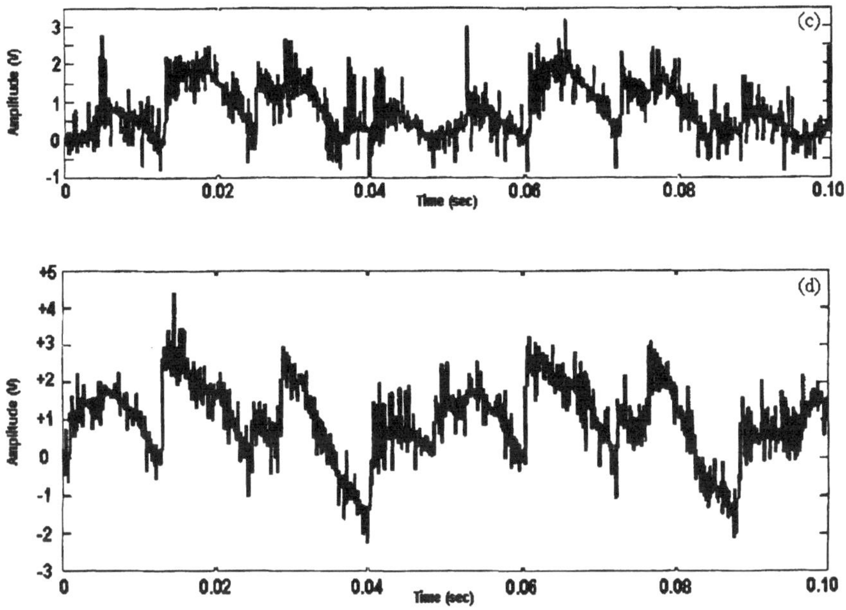

Fig. 4.6 (c) Plot of cutting force Fy *vs*. time for a new tool and (d) for a worn tool

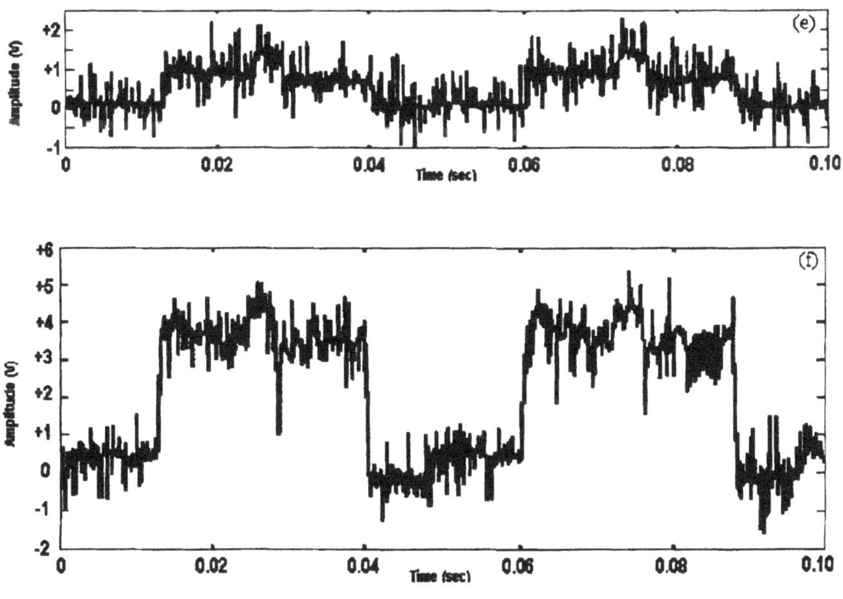

Fig. 4.6 (e) Plot of cutting force Fz *vs*. time for a new tool and (f) for a worn tool

In figures 4.6a-4.6f, the surges of the cutting forces due to the entry and exit of the individual cutting inserts are indicated by sudden changes of amplitude (each volt represents 100 N). It can also be shown directly that the amplitude and the variance of the cutting force signal are relevant to tool wear.

The spectra of the force signals were calculated in the frequency domain and figure 4.7 shows a typical set of spectra calculations for one specific set of cutting conditions. The cutting speed was 600 rpm, the feed rate was 0.4 mm/rev, and the cutting depth 0.5 mm. Ten frequency components noted (FS1, FS2, ..., FS10) were calculated by taking the average value in each frequency segment. The frequency components FS1 to FS5 evenly cover the frequency range from 0 Hz to 500 Hz, and FS6 to FS10 cover the frequency range from 500 Hz to 2500 Hz. Experimental results showed that frequency components in the first five frequency segments are of greater importance. The components in FS1 show very apparent variations when tool wear value increases and the components in FS2 to FS5 increase proportionally along with the development of tool wear. The amplitude and distribution patterns of the components in FS6 to FS10 have also noticeable variations when tool wear value increases.

4.6 DESIGN AND DEVELOPMENT OF A NEURO-FUZZY HYBRID PATTERN RECOGNITION SYSTEM

An experienced human machine tool operator can be very successful at monitoring tool wear states because of two major reasons. First, many types of information have been sensed and collected, including visual information (e.g. workpiece surface roughness, presence of smoke, form and colour of the chip), audio information (e.g. noise generated by the rubbing action of the tool flank on the workpiece), and other information (e.g. vibration amplitude on the machining structure, the smell of the smoke that represents the temperature of the cutting area). Secondly, the operator uses his knowledge and experience to associate all this information in order to obtain a correct recognition of the tool wear state. The operator's experience and knowledge about tool wear is acquired from a long period of machining practice.

As previously stated, four sensors have been used to collect information that represents the tool wear states from different aspects. The next step is to develop an intelligent data processing methodology to model the learning and decision making abilities of the human operator. Fuzzy logic and neural network pattern recognition techniques are effective means of integrating multisensor information and tool wear state recognition. By imitating the thinking and judgement modes of the human being, these techniques show some remarkable characteristics. No definite relationship between tool wear and sensor signal is necessarily required and faults caused by randomness and experimental noise can be tolerated to some extent.

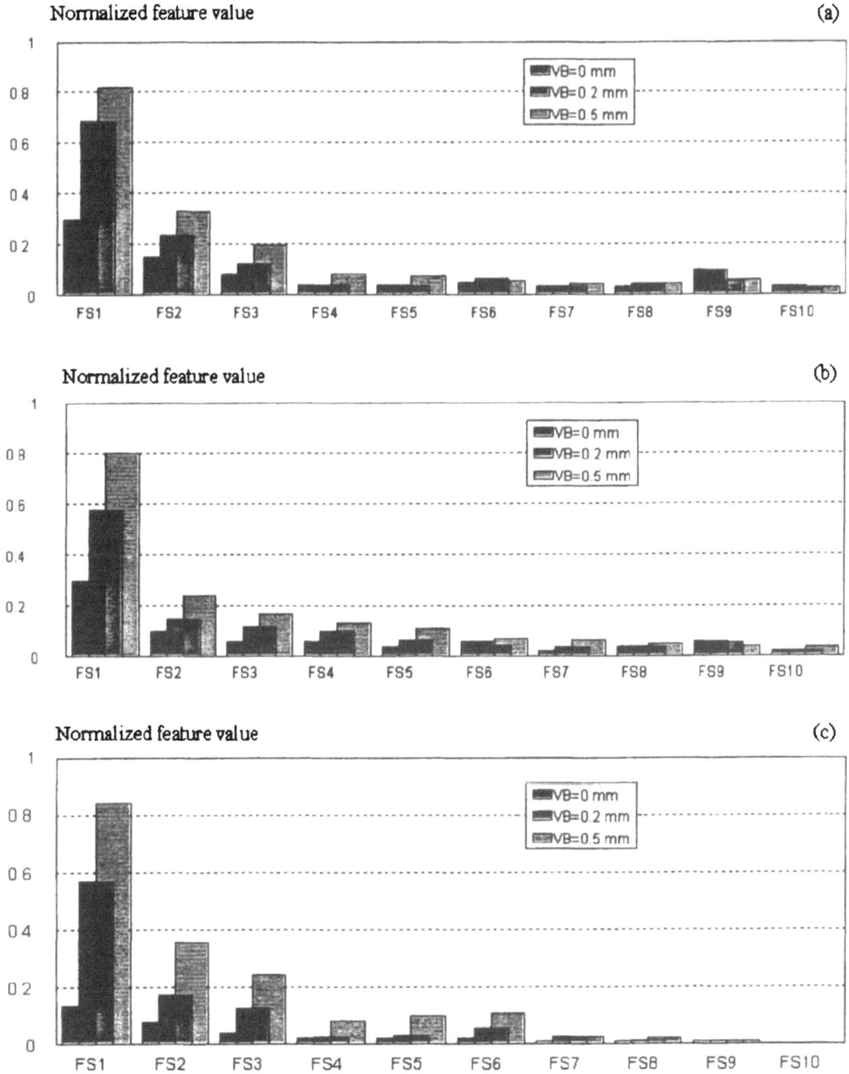

Fig. 4.7 The normalised frequency components of the cutting force signals (a) Fx, (b) Fy, (c) Fz, at different frequencies

4.6.1 Fuzzy Clustering Feature Filter

As presented in the following sections, the tool wear recognition process includes the fuzzification of signal features and the training of recognition networks. Four kinds of signals are used to describe tool condition comprehensively so that the tool wear recognition results can be more robust. But this comprehensiveness also means a comparatively large number of input features for the recognition networks. To improve the efficiency and reliability of the neuro-fuzzy pattern recognition algorithm, redundant features should be removed. It is well known that clustering techniques can be used to search for structures in data and in this study the fuzzy clustering technique was applied to develop an effective feature filter. A fuzzy clustering feature filter based on the FCM clustering methodology was developed to remove the redundant features (i.e. those features less relevant to tool wear). In the fuzzy clustering process, only a small number of features under each specific cutting condition were selected, but at the same time the confidence of correct recognition was improved. This process greatly improves the efficiency of the pattern recognition system and assures the high accuracy and reliability of the final results.

As previously indicated, for each of the models (i.e. tools that have standard wear values), fifty features were extracted from the sensor signals. Assuming there are *n* groups of repeated sampled feature values collected from the machining process for each model, each group of feature values can be treated as a target to be partitioned. Taking the features of all the models as the input of a fuzzy clustering system, those features should be assigned into the correct clusters (corresponding to different models) to which they belong. Those features whose variations of cluster centres for different models are smaller than a threshold ε_1 or whose variation of practical values corresponding to their own cluster centres are bigger than a threshold ε_2 should be removed. This is because they do not have recognisable variation along with the development of tool wear or they do not have stable values. In the practical process for filtering the signal features, the threshold values ε_1 and ε_2 are "floating". In order to effectively carry out the data fusion operation, a rule was set for the feature filtering action: for each sensor signal, the features remaining should be more than or equal to one and less than or equal to five. In the feature filtering process, the two threshold values were adjusted to meet the demands of the rule.

Of the fifty features originally extracted from the time and frequency domains the fuzzy clustering feature filter selects about twenty features, which are more relevant to tool wear state than the rest. These features used as input of the neuro-fuzzy pattern recognition network. The membership value indicates the confidence with which a target is assigned to the model to which it really belongs. This value will be much closer to unity after the redundant features have been removed. The performance of this repeated clustering process can be well controlled by adjusting the two thresholds. This has proved to be an effective feature filtering process. Redundant features are removed while at the same time the confidence of correct recognition is improved. In fact this is a pre-recognition process in which less and less signal features are used but the tool wear state classification can be seen as being successful. Consequently this also means that the size of the input space of the neuro-fuzzy recognition network is greatly reduced but its high accuracy and reliability can be assured.

4.6.2 Fuzzy Driven Neural Network

The tool wear state recognition can be realised by assigning a group of cutting inserts used in a machining process to a standard model with known flank wear value. Both the standard cutter and the cutter with an unknown tool wear value can be presented by features drawn from multisensor information. These features can be regarded as fuzzy sets in a power set. The degree of similarity between the corresponding feature pairs of the standard model and the target to be recognised can be calculated. By considering all the corresponding features, the target to be recognised can be assigned to the standard model to which it is most similar. So the key part of the tool wear recognition process is to calculate the degree of similarities of all of the corresponding feature pairs, or fuzzy sets, and to develop a methodology that can combine these fuzzy similarities according to their varying importance to make a correct estimation of the tool wear state. Treated as fuzzy sets, the signal features can have membership functions of triangular or normal distribution shapes. When comparing the degree of similarity of two fuzzy sets, the two quantity indexes used in this study are fuzzy closeness and fuzzy approaching degree. The former expresses how far away the two fuzzy sets are, and the latter is an indication of the degree of intersection of two fuzzy sets. Combining the two indexes generates a more representative fuzzy similarity measure, known as the two-dimensional fuzzy approaching degree. Here, artificial neural networks are used in a creative way to assign the signal features suitable weights and combine them to provide reliable tool wear state recognition results.

4. 6.3 Fuzzy Membership Function of Signal Features

The most important character of a fuzzy set is its membership function. It describes mathematically the ambiguity of the fuzzy set. The membership function $\mu(x)$ satisfies:

$$0 \leq \mu(x) \leq 1 \tag{4.1}$$

The membership function can be determined by using a fuzzy statistical method. If u_0 is an element in a power set U, i.e. $u_0 \in U$, n number of experiments are carried out to decide if u_0 belongs to a general set A^* in the power set U. So the membership degree of u_0 for the set A^* can be expressed as:

$$\mu(u_0) = \lim_{n \to \infty} \frac{\text{the number of } u_0 \in A^*}{n} \tag{4.2}$$

here n should be bigger than fifty. Along with an increase of n, $\mu(u_0)$ will tend to a certain value in [0,1] and this is the membership degree.

After a large number of milling operations had been carried out, it was determined that the features of the sensor signals in the tool wear monitoring process were normal distribution type fuzzy sets. For the signal features of the models (i.e. tools with standard wear values), considering that the values of the features vary in a limited range due to the influence of many factors, the membership function of the fuzzy set A_{ij} (j^{th} feature of the ith signal) can be expressed as:

$$A_{ij}(x) = \begin{cases} e^{-\frac{(x-a_{ij})^2}{2\sigma_{ij}^2}}, & a_{ij} - \sqrt{2}\sigma_{ij} \le x \le a_{ij} \\ 1, & a_{ij} \le x \le b_{ij} \\ e^{-\frac{(x-b_{ij})^2}{2\sigma_{ij}^2}}, & b_{ij} \le x \le b_{ij} + \sqrt{2}\sigma_{ij} \\ 0, & \text{others} \end{cases} \tag{4.3}$$

where σ is the standard deviation and [a, b] is the range of variation of the feature value. The shape of the membership function is represented by a solid line in figure 4.8.

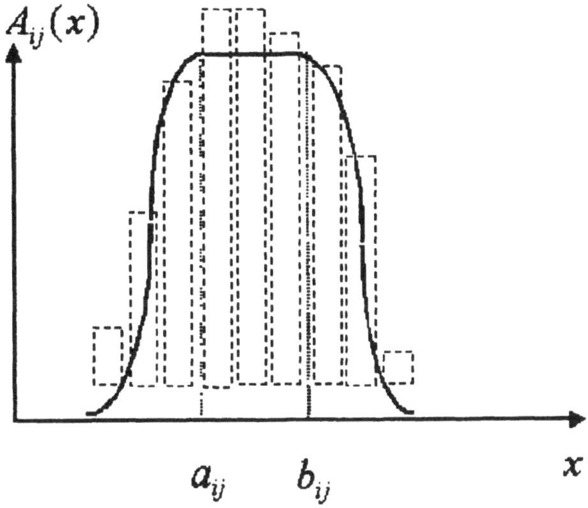

Fig. 4.8 Fuzzy membership function

The above assumption of the shape of the fuzzy membership was supported by a large number of experiments. The dotted line frames in figure 4.8 shows groups of experimental results. Sixty samples of cutting force F_x were collected from milling operations under the same cutting conditions using a cutter with flank wear, VB = 0.2 mm. The signal values varied from 365 Newton to 483 N and the nine frames in the figure are evenly placed in the whole range of the signal. The height of the frame is proportional to the number of sample values falling in the frame. In this case a = 398 N and b = 432 N. The pattern of the nine frames shows that the chosen modified normal distribution fuzzy membership function is suitable for representing the signal distribution property. Machining experiments for other signal features provided similar results [15]. To determine the values of the coefficients in the formula, several groups of cutters possessing standard wear values were used in milling operations. K groups of specimens are drawn for the jth feature, then for each group the mean value $\bar{x}_{ijt}$ and the standard deviation σ_{ijt} can be calculated (for t = 1, 2, ..., k). So a_{ij} and b_{ij} are set as the maximum and minimum values of $\bar{x}_{ijt}$ and σ_{ij}^2 takes the mean value of σ_{ijt}^2. For a certain group of inserts with unknown wear value, its jth feature can also be regarded as a normal distribution fuzzy set. It has the following membership function:

$$B_j(x) = e^{-\frac{(x-\bar{x}_j)^2}{2\sigma_j^2}} \qquad x_j - \sqrt{2}\sigma_j \le x \le x_j + \sqrt{2}\sigma_j = 0 \text{ for all other} \tag{4.4}$$

where $\bar{x}_j$ and σ_j are the mean value and standard deviation of the jth signal feature.

4.6.4 Fuzzy Closeness

For two fuzzy sets *A* and *B*, the membership functions can be defined as $y = A(x)$ and $y = B(x)$, where $y \in [0,1]$ and $x \in [0,1]$ when all the feature values have been normalised. Two corresponding inverse functions $x = A^{-1}(y)$ and $x = B^{-1}(y)$ can then be defined. *X* is a double solution function and here only the smaller half of the symmetry solution is adopted. The fuzzy closeness between two fuzzy sets is defined as follows:

when $y = \{y_1, y_2, \ldots, y_n\}$,

$$M_p(A,B) = 1 - \frac{1}{n}\left[\sum_{i=1}^{n} \left|A^{-1}(y_i) - B^{-1}(y_i)\right|^p\right]^{1/p} \tag{4.5}$$

when $y = [a,b]$,

$$M_p(A,B) = 1 - \frac{1}{b-a}\left[\int_a^b \left|A^{-1}(y) - B^{-1}(y)\right|^p dy\right]^{1/p} \tag{4.6}$$

where *p* is a positive real number and in the general situation it can take the value of 1 and $0 \le a \le b \le 1$. $M_p(A,B)$ is the fuzzy closeness between fuzzy set *A* and *B*. Fuzzy closeness is a novel quantitative index that represents the similarity of two fuzzy sets.

4.6.5 Fuzzy Approaching Degree

The fuzzy approaching degree is another quantity representing the similarity of two fuzzy sets. Assuming that F(X) is the fuzzy power set of *X* and the mapping, $N: F(X) \times F(X) \to [0,1]$, satisfies:

- $\forall A \in F(X), N(A,A) = 1$
- $\forall A, B \in F(X), N(A,B) = N(B,A)$
- if $A,B,C \in F(X)$ satisfies $|A(x) - C(x)| \geq |A(x) - B(x)| \ \forall x \in X)$ then $N(A,C) \leq N(A,B)$

So the map *N* is the grade of approaching degree in F(X) and N(A,B) is called the approaching degree of *A* and *B*. The approaching degree can be calculated using different methods and here the inner and outer product method is applied.

4.6.6 Fuzzy Approaching Degree Expressed by Inner and Outer Product

The fuzzy sets in a limited power set can be expressed as fuzzy vectors. Suppose *A* = (a_1, a_2, ..., a_n) and *B* = (b_1, b_2, ..., b_n), imitating the vector inner product calculation method used in linear algebra, the special product:

$$A \bullet B = \bigvee_{k=1}^{n} (a_k \wedge b_k) \tag{4.7}$$

is called the inner product of fuzzy sets *A* and *B*. Here, the symbols for multiplication "•" and addition "+" have been replaced by "$\wedge$" and "$\vee$". Extending the concept to the arbitrary power set *X*, suppose *A*, *B* ∈ F(X):

$$A \bullet B = \bigvee_{x \in X} (A(x) \wedge B(x)) \tag{4.8}$$

is called the inner product of fuzzy sets *A* and *B*. The dual part of the inner product is the outer product. Suppose *A*, *B* ∈ F(X):

$$A \oplus B = \bigwedge_{x \in X} (A(x) \vee B(x)) \tag{4.9}$$

is called the outer product of fuzzy sets *A* and *B*. The inner product and outer product have the following properties:

$$(A \bullet B)^c = A^c \oplus B^c, \ (A \oplus B)^c = A^c \bullet B^c \tag{4.10}$$

here *c* indicates the complementary calculation, i.e. $a^c = 1\text{-}a$,

$$A \bullet B = B \bullet A, \ A \oplus B = B \oplus A \tag{4.11}$$

$$A \subseteq B \Rightarrow A \bullet C \leq B \bullet C \text{ and } A \oplus C \leq B \oplus C \tag{4.12}$$

$$A \bullet A^c \leq \frac{1}{2}, \; A \oplus A^c \geq \frac{1}{2} \tag{4.13}$$

For $A \in F(X)$:

$$\bar{a} = \vee_{x \in X} A(x), \; \underline{a} = \wedge_{x \in X} A(x) \tag{4.14}$$

$\bar{a}$ and $\underline{a}$ are called the upper bound and lower bound of the fuzzy set,

$$A \bullet A = \bar{a}, \; A \oplus A = \underline{a} \tag{4.15}$$

$$A \bullet B \leq \bar{a} \wedge \bar{b}, \; A \oplus B \geq \underline{a} \vee \underline{b} \tag{4.16}$$

$$A \subseteq B \Rightarrow A \bullet B = \bar{a}, \; A \oplus B = \underline{b} \tag{4.17}$$

$$\lambda \in [0,1], \text{ then } (\lambda A) \bullet B = \lambda \wedge (A \bullet B) = A \bullet (\lambda B) \tag{4.18}$$

It can be seen from the properties stated above that the inner product and outer product will vary when fuzzy set *B* approaches *A*. When *A* and *B* are closest to each other, A•B reaches the highest value $\bar{a}$ and A⊕B has the lowest value $\underline{a}$. When *A* and *B* are opposite to each other, A•B is smaller than 0.5 and A⊕B is bigger than 0.5. This can be seen in figure 4.9.

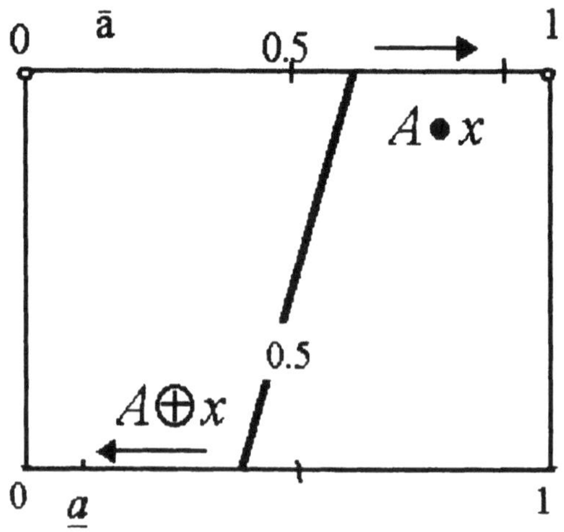

Fig. 4.9 Schematic representation showing that the dynamic picture {A•*x*, A⊕*x*} varies along with *x*

In the figure, the dynamic characteristic of x approaching A is a "polar" $\{A \bullet x, A \oplus x\}$. It rotates clockwise with the limitation of $\{\bar{a}, \underline{a}\}$. When x leaves A, the polar rotates anti-clockwise with the limitation of $\{0,1\}$. Let's define the approaching degree of fuzzy sets A and B as:

$F(X,1,0)=\{ A \mid A \in F(X),\ \bar{a} = 1,\ \underline{a} = 0 \}$ if $A, B \in F(X,1,0)$ then

$$(A, B) = (A \bullet B) \wedge (A \oplus B)^c \tag{4.19}$$

It is easy to calculate and quite effective for expressing the similarity degree of two fuzzy sets.

4.6.7 Two-dimensional Weighted Approaching Degree

Using the conventional fuzzy pattern recognition technique, fuzzy distances (such as approaching degree) between corresponding features of the target to be recognised and different models are calculated. Combining these distances can help determining the synthetic fuzzy distance between the target and different models. The target should be assigned to one of the models, which has the shortest fuzzy distance (or highest approaching degree) with it. Because most features have vague boundaries, using fuzzy membership functions to represent their characteristics and fuzzy distance to describe the similarity of corresponding features are quite appropriate. Fuzzy pattern recognition techniques are thus reliable and robust. These methods can be further improved by assigning suitable weights to different features in order to reflect their specific influences in the pattern recognition process. The two fuzzy similarity measures mentioned above can also be combined to describe the similarity of two fuzzy sets more comprehensively.

Approaching degree and fuzzy closeness reflects the similarity of two fuzzy sets from different aspects. For two fuzzy membership functions, approaching degree reflects the geometric position of the intersecting point and fuzzy closeness shows how close they are. Combining the two measures can provide a more accurate and extended index of similarity of two fuzzy sets. Here, the average value of the two measures is taken as a new and more representative fuzzy similarity index and is given the name of the two-dimensional fuzzy approaching degree. Two-dimensional approaching degrees between different features of the targets and that of the models also have changing importance in different cutting environments. In this study, artificial neural networks are employed to assign to the approaching degrees suitable weights to provide 2-D weighted approaching degrees. This makes the tool wear recognition process more accurate and reliable.

4.7 FUZZY DRIVEN NEURAL NETWORK

Artificial Neural Networks (ANNs) have the ability to recognise inputs. The weights between neurones are adjusted automatically in the learning process to minimise the difference between the desired and actual outputs. In this study, a back propagation ANN was connected with the fuzzy logic system to carry out tool wear recognition. The fuzzy driven neural network is shown in figure 4.10.

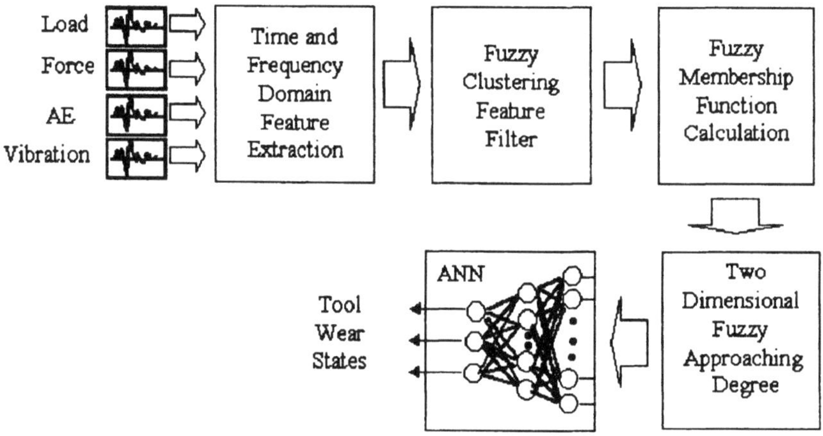

Fig. 4.10 Schematic of the fuzzy driven neural network

The 2-D approaching degree calculation results are used as inputs to the ANN. The associated weights can be updated as:

$$w_i(new) = w_i(old) + \alpha\delta\, x_i \qquad (4.20)$$

Here α, δ and x_i are the learning constant, associated error measure and input to the ith neurone. In this updating process, the ANN recognises the patterns of the features corresponding to a certain tool wear state. It can be considered that the ANN assigns to each input a proper synthesised weight and the ANN recognises the target by calculating the weighted approaching degree between the target and different models. This enables the recognition process to be more reliable. In the tool condition monitoring process, for each model, *N* groups of sensor signals were collected from the metal cutting process. After data processing in the time and frequency domain, signal features were extracted. Redundant features were removed by using the fuzzy clustering feature filter. Among those N groups of selected features, *K* groups (K = N-20) were used to establish the membership functions for each feature of each model. Ten groups were used for the training and another ten groups were used for the verification of the ANNs. Using equation (4.3) it is possible to determine the fuzzy membership function for all the features of each model. Therefore all the models are memorised by the monitoring system and a membership function library is established. These membership functions of the models are of varied normal distribution type. The membership functions of the twenty groups of training and verification features can then be calculated. These groups of features are used not as models but as targets having the same standard wear values as the corresponding models. The 2-D approaching degrees between the training and verification features and the corresponding features of different models can be selected using equations (4.5) and (4.19) as being the training and verification input. The training and verification targets are the corresponding models to which the inputs should belong.

Taking a typical training case as an example, for the milling operation under a specific set of cutting conditions, there were three models and fifty signal features. After the feature filtering process, 22 features remained. So the training input had the size of 22×30. It should be noticed that inputs to the ANN were not original signal feature values but 2-D fuzzy approaching degrees. The latter were processed results from the fuzzy logic system. This means the ANN had a lighter modelling load and higher training efficiency than ordinary pattern recognition ANN with original signal features as its input.

In the training process, for each model and under a specific set of cutting conditions, an artificial neural network was established that can determine whether the incoming targets belong to this model. After the training, the constructed frame and associated weights of the ANN can reflect the distinct importance of different features for each model under specific cutting conditions. In this way the future tool wear recognition results can be reliable and accurate. The determination of the membership functions of all the features for each model and the construction of ANNs for tool wear value recognition mark the end of the learning stage. Ten groups of verification inputs are then used to test the constructed ANN, and they should be assigned to correct models.

In a practical tool condition monitoring process, a tool with unknown wear value is the target that needs to be recognised. The membership functions of all the features of the target can be calculated first. The 2-D approaching degrees of the corresponding feature pairs of the standard models and the target to be recognised can be calculated next and become the inquiry input of the ANNs. Pre-trained ANNs are used to calculate the weighted approaching degree. Finally, the tool should be assigned to the model with which it has the highest weighted approaching degree. The process of applying the fuzzy driven neural network to accomplish tool wear states recognition can be divided into three steps:

- *Determination of membership functions.* I.e. fuzzy clustering feature filtering, tool wear relevant features are selected. For each model (standard cutters), the verified normal distribution type fuzzy membership functions of all the features are calculated and stored as the standard database. For the object (i.e. the cutter used in the machining process), normal distribution type membership functions of signal features are also determined.
- *Calculation of the two-dimensional fuzzy approaching degree.* The fuzzy approaching degrees and fuzzy closeness between the features of the object and the corresponding features of the models are calculated. Next, the two fuzzy indexes are combined to form the 2-D fuzzy approaching degree, which can more accurately represent distances between fuzzy sets.
- *Using the fuzzy driven neural networks to recognise tool wear states.* Networks are first trained by using the information of the models so they are capable to recognise signal patterns corresponding to different flank area widths under specific cutting conditions. In the practical tool condition monitoring process, the trained networks can recognise tool wear states according to the incoming signal feature vectors.

4.7.1 Determination of Fuzzy Membership Functions

The signal features of the models are treated as normal distribution type fuzzy sets. The membership function of the fuzzy set A_{ij} (j^{th} feature of the ith signal) can be expressed as shown previously in equation (4.3), where σ is the standard deviation and [a, b] is the range of variation of the feature value.

For each sensor signal feature under a certain cutting condition, thirty samples of signal collected from the machining process are used to determine the fuzzy membership. This forms three groups of samples when every group contains ten data samples. Then for each group the mean value $\bar{x}_{ijt}$ and the standard deviation σ_{ijt} can be calculated for t = 1, 2, 3. So a_{ij} and b_{ij} can be set as the maximum and minimum values of $\bar{x}_{ijt}$ and σ_{ij}^2 can take the mean value of σ_{ijt}^2. For a certain group of inserts with unknown wear value, its jth feature can be regarded as a normal distribution type fuzzy set. As shown previously it has the membership function given by equation (4.4), where $\bar{x}_j$ and σ_j are the mean value and standard deviation of the jth signal feature. For each sensor signal feature under certain cutting conditions, ten samples of signal are collected from the machining process are used to determine the parameters.

4.7.2 Fuzzy Similarity

Fuzzy Approaching Degree

Fuzzy approaching degree is an index expressing how close the two fuzzy sets are. As previously described, inner and outer products are used to determine the approaching degree. So if $A = (a_1, a_2, \ldots, a_n)$ and $B = (b_1, b_2, \ldots, b_n)$, are fuzzy sets, the inner and outer products of fuzzy sets A and B respectively can be calculated from equations (4.8) and (4.9). The approaching degree of A and B can be calculated using equation (4.19). Suppose A, $B \in F(X)$ and A, B are normal distribution fuzzy sets with membership functions as follows:

$$A(x) = e^{-(\frac{x-a}{\sigma_1})^2}, \quad B(x) = e^{-(\frac{x-b}{\sigma_2})^2} \qquad (4.21)$$

From figure 4.11 it can be seen that A•B is the height of $A \cap B$, or the peak value of the membership function curve. The membership function curve of $A \cap B$ can be considered as the lower part of the membership function curves of A and B. It is the curve CDE in figure 4.11. Hence its peak value is x^*, which is the co-ordinate of the intersection point of the membership function curve A(x) and B(x).

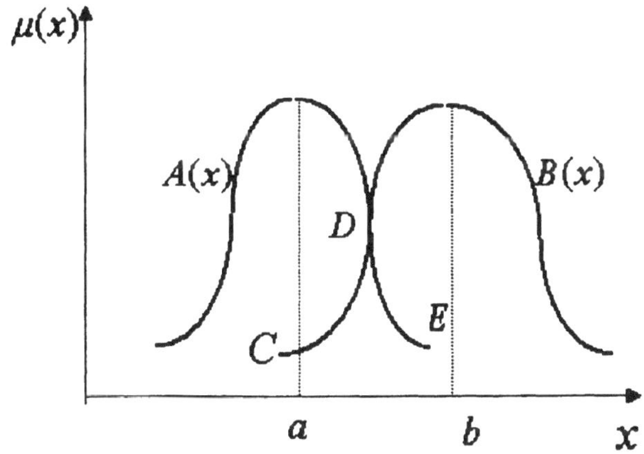

Fig. 4.11 Representation of two normal distribution membership function curves

Let's write:

$$(\frac{x-a}{\sigma_1})^2 = (\frac{x-b}{\sigma_2})^2$$

so:

$$x^* = \frac{a\sigma_2 + b\sigma_1}{\sigma_1 + \sigma_2} \quad \text{and} \quad A \bullet B = e^{-\left(\frac{b-a}{\sigma_1+\sigma_2}\right)^2}$$

Similarly, as:

$$\lim_{x \to \infty} A(x) = \lim_{x \to \infty} B(x) = 0, \text{ so: } A \oplus B = 0.$$

Thus, the approaching degree of A and B can be expressed as:

$$(A, B) = (A \bullet B) \wedge (A \oplus B)^c = e^{-(\frac{b-a}{\sigma_1+\sigma_2})^2} \tag{4.22}$$

Fuzzy Closeness

Fuzzy closeness is a novel index for describing how close are the two fuzzy sets [15]. It shows the similarity of two fuzzy sets from a different angle compared with the fuzzy approaching degree. For two fuzzy sets *A* and *B*, when all the feature values have been normalised, the membership functions can be defined as y = A(x) and y = B(x), where $y \in [0,1]$ and $x \in [0,1]$. Two corresponding inverse functions $x = A^{-1}(y)$ and $x = B^{-1}(y)$ can then be defined. Considering ten corresponding point pairs of the two membership functions, the fuzzy closeness between two fuzzy sets is defined as follows. When $y = \{y_1, y_2, \ldots, y_{10}\}$,

$$M_p(A, B) = 1 - \frac{1}{10}\left[\sum_{i=1}^{10} \left|A^{-1}(y_i) - B^{-1}(y_i)\right|^p\right]^{1/p} \tag{4.23}$$

where *p* is a positive real number and in the general situation it can take the value of 1. $M_p(A, B)$ is the fuzzy closeness between fuzzy sets *A* and *B*. Fuzzy closeness is another quantitative index that represents the similarity of two fuzzy sets.

Two-Dimensional Fuzzy Approaching Degree

Fuzzy approaching degree and fuzzy closeness describe the fuzzy similarity from different aspects. The mean value of the two quantitative indexes, given the name 2-D fuzzy approaching degree, is defined as a new fuzzy measure. This one can more comprehensively express the fuzzy distance between two fuzzy sets than in the conventional fuzzy pattern recognition system where only one index (e.g. approaching degree) is employed.

4.7.3 Establishment of the Fuzzy Driven Neural Network

Using fuzzy sets to express signal features in the tool condition monitoring process is very appropriate as these features do have vague boundaries. The developed 2-D fuzzy approaching degree is also a reliable quantitative index expressing the similarity of two fuzzy sets. But in the final stage of the conventional fuzzy pattern recognition process, the recognition stage, all the signal features are treated as equally important. This is not very reasonable because the tool wear relevant degree of each feature always changes under different machining conditions.

It is obvious that granting all features with the same weight (of 1) is a main weak point of the traditional fuzzy pattern recognition process. But the majority of the weight determining methods that have already been put forward are not very practical. In this study, a unique feature weight assigning process has been developed. The 2-D fuzzy approaching degree values are used to represent the fuzzy similarity of each feature pair. But at the decision making stage, artificial neural networks are used to integrate the fuzzy similarity results, assign each feature a synthetic weight and provide reliable predictions of the tool wear values. The capability of the ANNs to automatically update the weights to reach optimised performance is fully utilised and the reliability and accuracy of the tool wear monitoring system can be greatly improved. In fact in the tool wear value recognition process, an ANN takes the amplitude of all the 2-D approaching degrees into account. But, it also considers the distribution patterns of those approaching degrees for different models to make the final decision of the wear value. In thc training process, an ANN can adjust the importance of each feature by adjusting weights to obtain the best performance. In the recognition process, the trained ANN will assign each feature a suitable, synthetic weight and then integrate approaching degrees of all the feature pairs to provide a tool wear value recognition result that is more reliable than that of a traditional fuzzy recognition system. Figure 4.12 shows the calculation process of the fuzzy driven neural network.

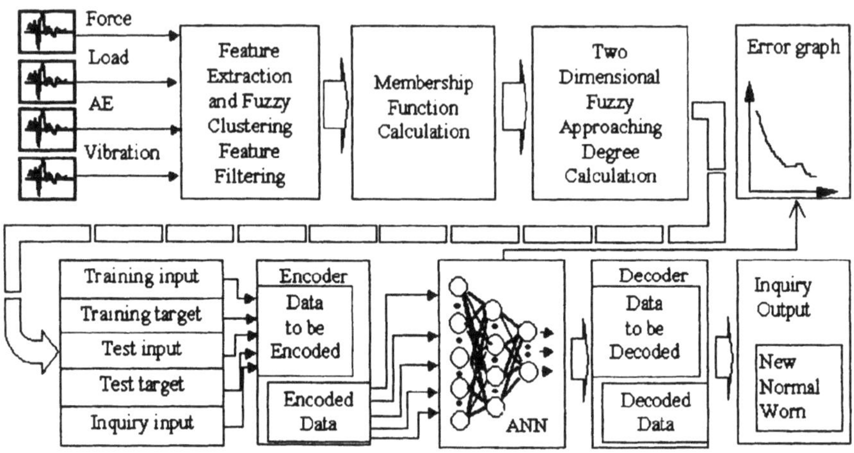

Fig. 4.12 Schematic application of the fuzzy driven neural network

The training process of the ANNs is as follows. First, for every model under a specific set of cutting conditions, fifty groups of sensor signals are collected from the practical machining process and the same number of groups of signal features can be obtained after the feature extracting and filtering process. Using equation (4.3) to process thirty groups of data, one can determine the varied normal distribution type membership functions of all the selected features and establish a model. The remaining twenty groups of data are processed to obtain twenty groups of normal distribution type membership functions. The centre of the membership function of a feature is the feature value and the standard deviation value can be calculated using ten groups of feature values. These twenty groups of features can be considered as sub-models of each model having the same standard tool wear values. Next, the 2-D approaching degree between the corresponding features of ten sub-models and six models are calculated. The results can be used as the training and testing inputs of an ANN. The ANN integrates the input and records the corresponding patterns. The sub-models should only belong to the corresponding main models.

So, under a specific cutting condition, assuming there are n models, n fuzzy driven ANNs will be trained. This kind of small size neural network for each model only needs to represent the input/output mapping in a comparatively small range, therefore having high efficiency and reliability. In most cases, the training time for a network is less than five minutes. This highly efficient training process, which also benefits greatly from the input space reduction realised by the feature filtering operation, means that it is really possible for the developed tool condition monitoring system to be applied in industrial production. After training the structure and associated weights of each ANN reflect the features of a model under a specific set of cutting conditions. The neural network can decide whether the incoming object belongs to this very model. The other ten sub-models are then used to test the trained ANN. The determination of the membership functions of all the selected features for each model and the construction of ANNs for recognition mark the end of the learning stage.

In a practical tool condition monitoring process, the tool with unknown wear value is the object that will be recognised. The membership functions of all the features of the object can be determined using equation (4.4) to process ten groups of selected signal features. The 2-D approaching degree of the corresponding features of each standard model and the object can be calculated, and become the inquiry input of the ANNs. The ANNs then determine the degree to which the target belongs to the different models. Finally, the target should belong to the model with which it has the highest dividing confidence. Here, the dividing confidence of a fuzzy driven ANN is defined as the ratio of the real output of the corresponding output node and the ideal output. The recognition confidence of a target is the value of the difference of its highest and second highest dividing confidence for two models.

4.8 APPLYING FUZZY DRIVEN NEURAL NETWORKS UNDER COMPREHENSIVE CUTTING CONDITIONS

In order to verify the performance of the tool condition monitoring system, including the intelligent pattern recognition algorithm, a large number of milling experiments have been completed [15]. The experiments were designed to be representative of the performance of the monitoring system under the majority of popular cutting conditions. This was achieved using an orthogonal table which included four cutting parameters, i.e. cutting speed, cutting depth, feed rate and workpiece material. An 80 mm diameter face mill cutter with six inserts was used in the milling operations and the tool wear to be recognised was divided into three groups, 0.0 mm, 0.2 mm and 0.5 mm. All the targets were correctly classified to the appropriate tool wear group and the average recognition confidence for each target was approximately 80%. The results showed that fuzzy driven neural networks are quite effective and reliable for estimating tool wear values.

4.9 CONCLUSION

A tool wear monitoring methodology suitable for milling operation has been developed. The methodology is composed of three major parts, i.e. signal collection and conversion, feature extraction and intelligent pattern recognition. Based on this, a practical tool wear monitoring system has been established. A large number of experiments have been carried out to evaluate the tool wear monitoring methodology and the practical monitoring system. It was demonstrated that load, vibration, AE and cutting force sensors are applicable for monitoring tool wear in milling operations. The sensors are relatively robust and can survive in the harsh working environment of an intermittent machining process such as milling. These sensors detect signal variations at different working frequency ranges. Loss of sensitivity of one sensor under a specific cutting condition can be compensated by the others. Experimental results have shown that the fifty features extracted from the time and frequency domains are relevant to changes of tool wear state. This makes accurate and reliable pattern recognition possible. In the time domain, mean value, standard deviation, histogram component, kurtosis are calculated and found to be relevant to the tool wear value. In the frequency domain, the mean amplitude in some frequency ranges and the distribution pattern of the power spectrum of the cutting force signal can also be related to changes of the tool wear states.

The fuzzy signal processing process is effective and reliable. Taking a normal distribution function as the fuzzy membership function for the features drawn from sensor signals agrees well with the actual situation. The 2-D fuzzy approaching degree is a reliable index for presenting fuzzy similarity of signal feature pairs. This process has the significant advantages of being suitable for different machining environments, is robust to noise and is tolerant to faults. The fuzzy driven neural network is a very effective pattern recognition system. The combination of ANN and fuzzy logic systems integrates the strong learning and classification ability of the former and the flexibility of the latter to express the distribution characteristics of signal features with vague boundaries. This methodology indirectly solves the weight assignment problem of the conventional fuzzy pattern recognition system and the resulting fuzzy driven neural network has greater representative power, higher training speed and is more robust. This unique pattern recognition system can estimate tool wear values accurately and reliably under a wide range of cutting conditions.

Equipped with an advanced pattern recognition methodology, the established intelligent tool condition monitoring system meets the major demands for modern condition monitoring applications.

REFERENCES

1. Leem C.S., Dornfeld D.A., Dreyfus S.E., *A customised neural network for sensor fusion in on-line monitoring of tool wear*, Trans. ASME Journal of Eng. for Industry, 1995, 117(2):152-159.
2. Papazafiriou T., Elbestawi M.A., *Development of a wear related feature for tool condition monitoring in milling*, Proc. of the USA-Japan Symp. on Flexible Automation, 1988, Minneapolis, USA, 1009-1016.
3. Rangwala S., Dornfeld D.A., *Integration of sensors via neural networks for detection of tool wear states*, Symp. on Integrated and Intelligent Manufacturing Analysis and Synthesis, 1987, Pittsburgh, USA, 109-120.
4. Rangwala S., Dornfeld D.A., *Sensor integration using neural networks for intelligent tool condition monitoring*, Trans. ASME Journal of Eng. for Industry, 1990, 112(3):219-228.

5. Fu P., Hope A.D., King G.A., *Tool wear state recognition using a neurofuzzy classification architecture*, Insight, 1998, 40(8):544-547.
6. Sokolowski A., Dornfeld D.A., *On designing tool wear monitoring systems using supervised and unsupervised neural networks*, Proc. Joint Hungarian-British Mechatronics Conf., 1994, Budapest, Hungary,17-22.
7. Rao B.K.N., Hope A.D., Wang Z., *Application of Walsh spectrum analysis to milling tool wear monitoring*, Measurement and Control, 1998, 31(11):286-293.
8. Li X.L., Yao Y.X., Yuan Z.J., *On-line tool condition monitoring with an improved fuzzy neural network*, High Tech. Letters, 1997, 3(1):30-33.
9. Smith G. T., *CNC Machining Technology*, Springer-Verlag, 1993.
10. Du R., Elbestawi M.A., Wu S.M., *Automated monitoring of manufacturing processes Part 2: Applications*, Trans. of ASME J. of Eng. for Industry, 1995, 117(2):133-141.
11. Littlefair G., Javed M.A., Smith G.T., *Fusion of integrated multisensor data for tool wear monitoring*, Proc. of IEEE Int. Conf. on Neural Networks, 1995, Perth, Australia, 734-737.
12. Bezdek J.C., *Pattern Recognition with Fuzzy Objective Function Algorithms*, Plenum Press, 1981.
13. Zeng L., Wang H.P., *Machine fault classification: a fuzzy set approach*, Int. J. Adv. Manufac. Tech., 1991, 6:83-94.
14. Ko T.J., Cho D.W., *Tool wear monitoring in diamond turning by fuzzy pattern recognition*, Trans. ASME J. of Eng. for Industry, 1994, 116(2):225-238.
15. Fu P., *An intelligent cutting tool condition monitoring system for milling operation*, PhD Thesis, 2000, Nottingham Trent University / Southampton Institute.

5 FUSION OF ULTRASONIC IMAGES FOR 3-D RECONSTRUCTION OF INTERNAL FLAWS

Xavier E. GROS[1], Liu ZHENG[2]
[1]Independent NDT Centre,
Bruges, France
[2]Kyoto University,
Kyoto, Japan

5.1 INTRODUCTION

Due to more stringent regulations and safety standards, in-service examinations of structures and materials in nuclear power stations have to be performed more thoroughly [1-3]. In this chapter, a case study related to the application of NDT data fusion following ultrasonic examinations of a Hungarian nuclear power station is described.

Several NDT methods are used for structural integrity assessment of Paks' nuclear power plant in Hungary. These include visual inspection (e.g. naked eye, camera, boroscope), ultrasound, radiography and acoustic emission. However, the ultrasonic multiple sensor system used provides a large amount of data in a format difficult to interpret. Some of the problems encountered with such systems have already been described in the literature and include inability to correctly interpret C and D-scans, incorrect defect sizing and inaccurate estimation of defect location [4]. Thus, there is a need for an efficient data fusion process that would facilitate defect characterisation and location, and reduce ambiguity.

The fact that three-dimensional (3-D) imaging techniques play an essential role in materials science has already been discussed in the literature [5]. On the other hand, it is shown in this chapter that, combined with data fusion, three-dimensional computer graphics visualisation is a useful tool to produce images that contain quantitative parameters that may not be directly available on original 2-D images. The idea of displaying defects in a 3-D manner has generated lots of interest in industry. Indeed, several benefits may result from 3-D defect and flaw reconstruction, with enhanced defect location and sizing the most obvious. It may also improve manufacturing process by identifying the shape and cause of a flaw, reveal hidden artefacts, or provide complementary information to help fracture engineers estimate the remaining life of a structure. Research is being carried out to combine information gathered from multiple NDT apparatus to generate a 3-D view of a structure or flaws. Very often industry aims to create a 3-D image based on 2-D images produced using one NDT technique but in different operating conditions, or with different sensors.

Because of its popularity, ultrasonic testing was among the first method to benefit from computer enhanced inspection technology [6]. Already in 1995, a 3-D ultrasonic testing system that uses computer graphics technology has been developed to detect and estimate defects in complex structures with curved surfaces and non-uniform thickness [7]. All data is computer controlled by a computer graphic workstation, and ultrasonic output results are superimposed on a CAD image of the previously built inspected structure. More recently, Stolk demonstrated that a tomographic approach may not be necessary to generate a 3-D image based on information contained in an ultrasonic signal only [8]. Several mathematical interpolations and approximations were necessary to reconstruct a 3-D image that contained sufficient information to clearly distinguish flaws in a tube.

Several techniques of image reconstruction have been used to generate a 3-D image of a defect or a structure. For example, Pastorino recovered the shape of a structure from a limited number of views obtained with microwave, computer tomography, radiography and ultrasound [9]. An algorithm to correlate ground penetrating radar signals has been developed to estimate the three co-ordinates, and thus exact position, of a buried object [10]. In the case of buried land mines, there is a clear need for accurate, easy to interpret 3-D NDT images. Maximum correlation between consecutive signals performed on A and B-scans from GPR inspections was performed to generate a 3-D plot accompanied with statistical uncertainty levels [10]. Capineri *et al.* estimated the location of mines by calculating the centre of gravity of the B-scan points distribution [10]. Summa *et al.* developed an interactive 3-D visualisation and image analysis package for NDI data processing [11]. Their system allows import, rendering, and manipulation of full waveform data from several commercial ultrasonic systems. They showed that 3-D visualisation can provide a relatively simple means for analysis of massive amounts of data without requiring arbitrary decimation of the data. Volume rendering makes accessible view angles not normally reachable from the outside, and can be useful to identify features that may not be readily apparent on a set of sequential 2-D views because of their relative size, location or orientation. Suzuki *et al.* developed a 3-D acoustic image processing technique for inspecting in-vessel structure in a sodium environment [12]. Another application that could benefit from 3-D graphic visualisation is underwater inspection of offshore structures and pipelines, for example by combining visual and acoustic information gathered with a ROV [13].

Gautier *et al.* discussed three strategies for the fusion of X-ray radiographs and ultrasonic data [14]. A decentralised architecture (e.g. deduce the reconstruction from results made separately for each set of data), a cascade architecture (e.g. to process one data set and account for both the result of the first processing and the second data set), and a centralised architecture for which all data is processed simultaneously. The centralised approach is more attractive since it is robust towards poor geometrical positioning of the data. This use of data fusion by EDF provides information in lateral direction from radiographs together with information from the vertical direction obtained with ultrasound. The difficulty associated with ultrasonic examination of stainless steel is difficult due to anisotropy and heterogeneous structure of the weld, and led Schmitz *et al.* to develop a ray tracing algorithm to follow longitudinal, horizontal and vertical polarised shear wave propagation from the base material through the cladding in three dimensions [15]. The algorithm was implemented in the 3-D-Ray-SAFT software that allows experimental modelling to visualise rays in 3-D and include reflections from arbitrarily placed flaws.

In civil engineering, the signal resulting from ultrasonic inspection of concrete is often noisy and the waves are highly scattered. Reconstruction of 3-D SAFT images is now possible through the combination of laser interferometric detection of ultrasonic signals with pulse compression and random speckle modulation techniques [16]. This improves the signal-to-noise (S/N) ratio and leads to better defect visualisation. Still for civil engineering applications. Prémel *et al.* proposed an eddy current tomographic technique by building a spatial map from measurements of a scattered magnetic field [17].

A common problem in NDT is the limited number of views (or incomplete information) available about a structure. Reconstructing operations using Gibbs statistics allowed image reconstruction from a limited number of computer tomography (CT) projections [18]. Not only can such operations improve signal interpretation, they may also be a solution to reduce long inspection time often associated with CT. Vengrinovich *et al.* reconstructed 3-D images of flaws from a limited number of cone-beam X-ray projections using a Bayesian reconstruction with Gibbs prior in the form of mechanical models such as non-causal Markov fields, and investigated the influence of noise on the reconstruction results [19-20]. They showed that increasing noise level subsequently contaminates the reconstructed image and that the use of wavelet transforms causes a decrease in image restoration [19]. They also showed that the inverse radon transform used for 3-D image reconstruction is not suitable in the case of a limited number of views. Because of the insufficiently filled radon space, data interpolation cannot be performed [20]. They used the Tichonov-Miller regularisation procedure to solve this problem [21-22], and the maximum entropy method to extract physical quantities from noisy data sets [23]. They demonstrated that with only four projections, a reconstructed 3-D image containing clear information can be produced, and claimed that the results of their proposed multi-step reconstruction procedure are comparable to a CT image that would otherwise require a larger number of projections to achieve an equivalent image reconstruction.

Because microscopy techniques have poor depth of resolution, and because some specimen features can only be measured in 3-D (e.g. spatial distribution, connectivity, size distribution), scanning microscopy approaches have to be sought [24]. Some alternatives consist in scanning sections of 2-D images through different focal planes and then combines them to obtain a 3-D image. Combining views becomes complicated when working with different angle projections. Knowledge about the true (or measured) 3-D shape of a defect is desirable for accurate structural assessment as such views give a more consistent image of defects contained in an object. Another alternative to reveal multiple features of a material, is to fuse microscope images obtained with different configurations [25]. In such applications, a registration problem would have to be solved, for example by alignment using an algorithm to match points in digitally captured microscope images.

5.2 DESCRIPTION OF THE PROBLEM

Ultrasonic inspection of nuclear reactor vessels, together with radiography and eddy current testing, are regularly performed for maintenance and safety reasons. Automated inspection systems made of multiple NDT sensors are currently used, but the signals from each sensor are presented individually, providing only limited information about the environment in which they operate.

Ultrasonic images used for this data fusion approach are actual (not simulated) images obtained using a commercial ultrasonic system currently in use for the inspection of a VVER-4 type reactor in Hungary. Ultrasonic examination of the weld was performed using an industrial ultrasonic testing system developed by Tecnatom from Spain. The probe of this apparatus is made of multiple ultrasonic sensors: 0, 35, 45, 60 and 70 degree angles. Figure 5.1 shows a schematic of the probe. Advantages of using multiple sensor systems are well known and include coverage of large areas, reduced inspection time, and - in certain cases - possibility for automation and remotely controlled inspections. Despite these advantages, the signal output of most multiple sensor NDT systems is equal in number to the number of sensors. In the case of identical sensors their signals can be useful for an accurate geographic location of a flaw. If dissimilar sensors are used, the information displayed from each individual sensor may make signal interpretation difficult [26].

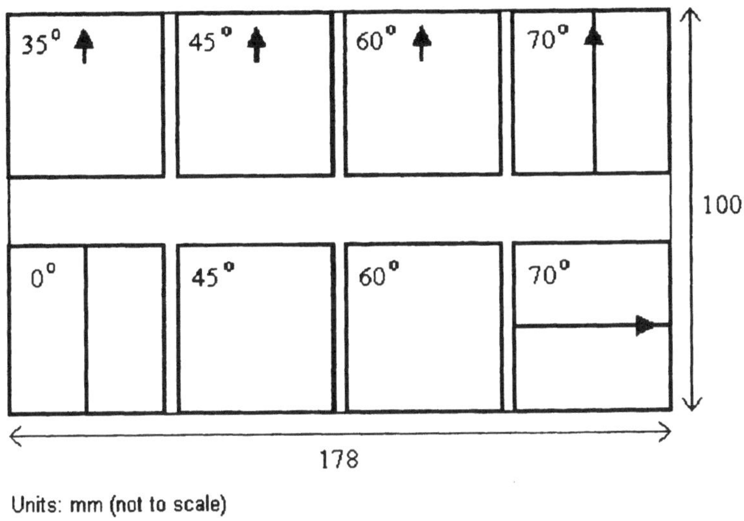

Fig. 5.1 Schematic of the ultrasonic probe configuration

One limitation of this multisensor equipment is in the display of ultrasonic inspection results. Indeed, the output of C and D scans are represented in the form of colour coded lines (figure 5.2). The four defects marked A, B, C and D indicated in figure 5.2 correspond to the defects on which the data fusion operation was performed. Although defects can be located on the images, sizing remains difficult and determination of their exact location very awkward. Additionally, a B-scan would be useful to determine the third dimension of the flaws (principally along the Y axis), but was not available. A B-scan is an ultrasonic projection of a defect along the Y and Z axes, a C-scan is a XY projection, and a D-scan a XZ projection. For clearer understanding, figure 5.3 presents a schematic of a material with a centreline weld containing an internal defect, and its B, C and D-scan projections.

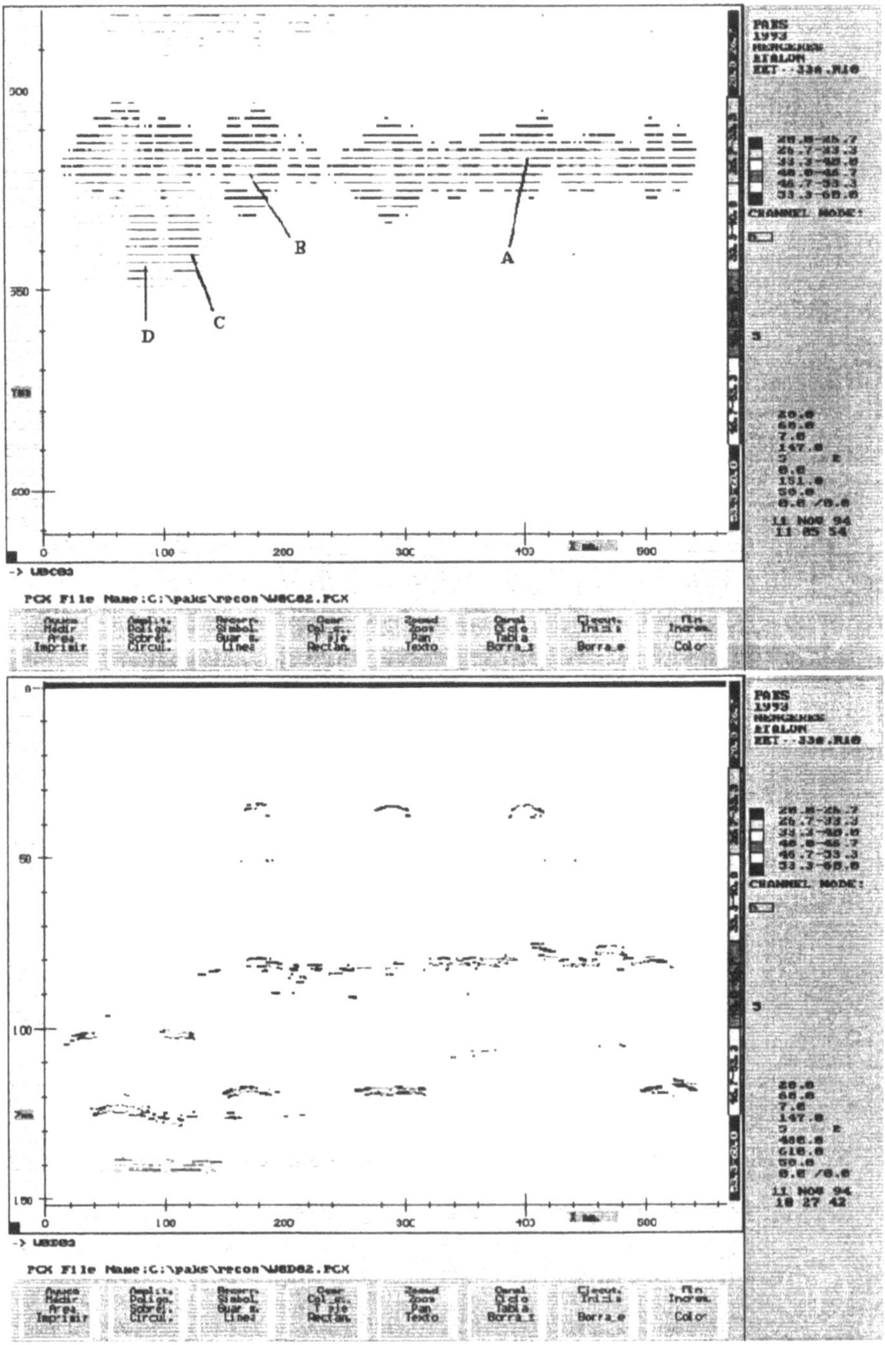

Fig. 5.2 Ultrasonic C-scan (top) and D-scan (bottom) of the weld from figure 5.4

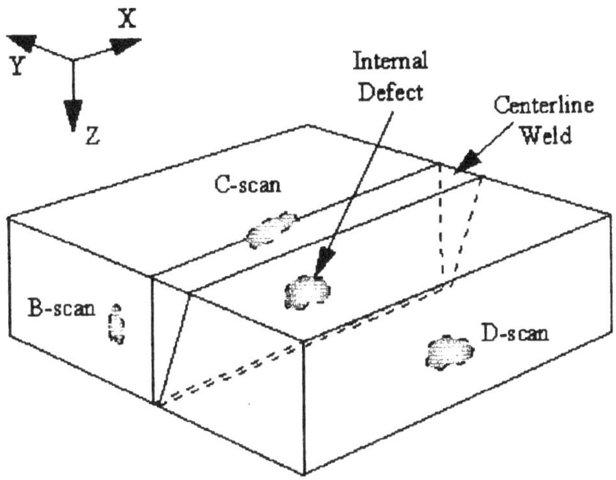

Fig. 5.3 Schematic of a material with a centreline weld containing an internal defect, and its B, C and D-scan projections

The part tested was a 150 mm thick and 565 mm long centreline weld on a rectangular shaped steel block (1040 × 565 × 150 mm) as shown in figure 5.5. The weld contains 13 simulated defects of different shapes and dimensions (table 5.1).

Table 5.1 Defect types and dimensions (numbers in brackets indicate the number of defects of a particular type or size)

Defect types	Dimensions (mm)
Spherical (5)	6 (2), 8 (2) and 10 (1) diameter
Flat (6)	3 (2), 5 (1), 7 (2) and 10 (1) long
Root failure (1)	-
Long failure (1)	20 length × 2 diameter

The possibility of NDT data fusion to combine information from multiple ultrasonic sensors, of identical or different angle, to enhance defect characterisation is presented in the next section.

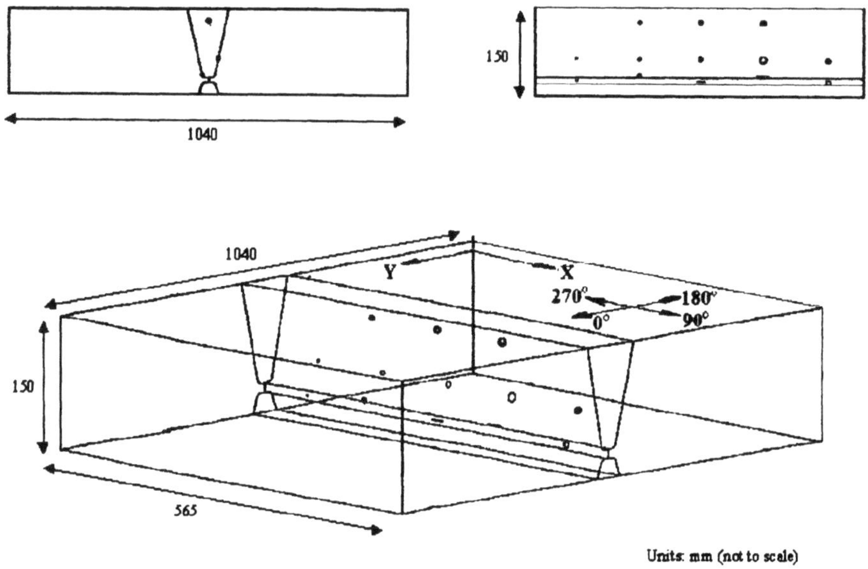

Fig. 5.4 Multiple views of the tested material containing artificial defects

5.3 FUSION OF ULTRASONIC C AND D SCANS FOR 3-D DEFECT RECONSTRUCTION

The analysis of C-scan and D-scan images requires registration of one or more points given in a 3-D co-ordinates system (x_i, y_i, z_i). Because of the complexity of the images (figure 5.2), the first step was to concentrate on a single defect and manipulate C and D scan views in such a manner that a 3-D image could be reconstructed that would display the defect in its entirety. This phase was followed by validation of the fusion operation on other defects. At first, we concentrated our work on a flat defect marked C in figure 5.2, and extended our concern to three additional flaws marked A, B (spherical), and D (flat) in figure 5.2. A defect displayed in a 2-D plane is in fact part of an object that also exists in 3-D co-ordinates. To facilitate image registration and reduce computer time calculations, the defect was extracted from the original C and D scans (figure 5.2). From the C-scan, one is able to determine the length and width (x and y dimensions) of the defect. The D-scan provides the width and thickness (x and z dimensions) of the defect. A simple mathematical translation was performed based on the assumption of symmetry. We are aware that the shape of the defect may be irregular and non-symmetric, but for a first approach this option was selected. All the calculations and graphical visualisations were performed using commercial software such as AVS and Matlab. Figure 5.5 shows the AVS network used to combine information from multiple images.

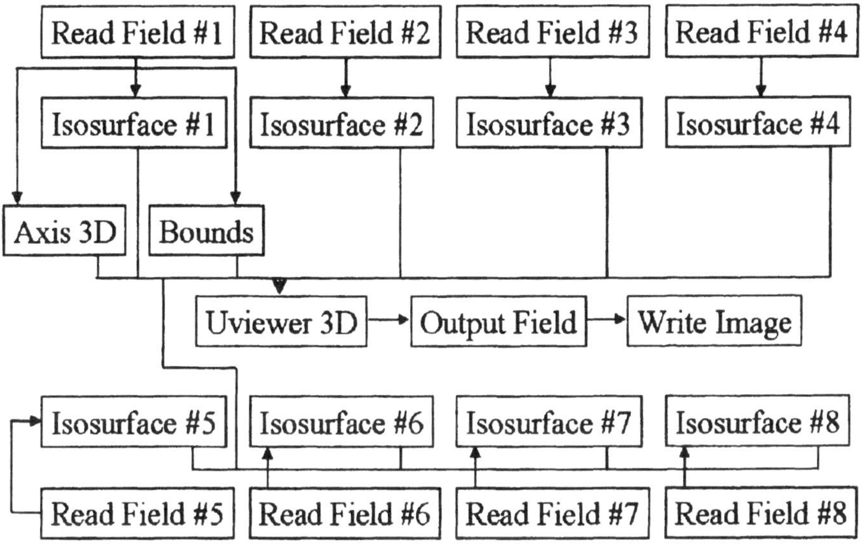

Fig. 5.5 AVS network used for the fusion of C and D scans

In order to combine C and D scan images, it was necessary to pre-process the original discrete signal into a more useful form. Figure 5.6 indicates the different image processing phases performed. These include image separation based on RGB values of the original line-image, a Gaussian blurring function followed by thresholding, and finally edge finding using a Canny operator that helped to delineate the defect. The outcome of this data fusion operation led to the generation of 3-D computer generated shapes of the defects considered (figure 5.7)

Finally, combination of both the defect and a CAD representation of the actual inspected material and weld was carried out. The result of this operation is a skeleton of the material, together with a defect as shown in figure 5.8. Such images can be rotated using a computer mouse positioned directly on the image, thus allowing a more accurate defect location. However, because of pre-processing reasons and of the small size of the flaw, the latter was not represented to scale. Nevertheless, this does not affect defect location and sizing. Indeed, sizing is performed on a separate window displaying the defect to its real scale, if the operator either clicks with the mouse on the defect or uses the zoom option of the software on a defect. If there is more than one defect, more windows can be opened (i.e. one for each defect).

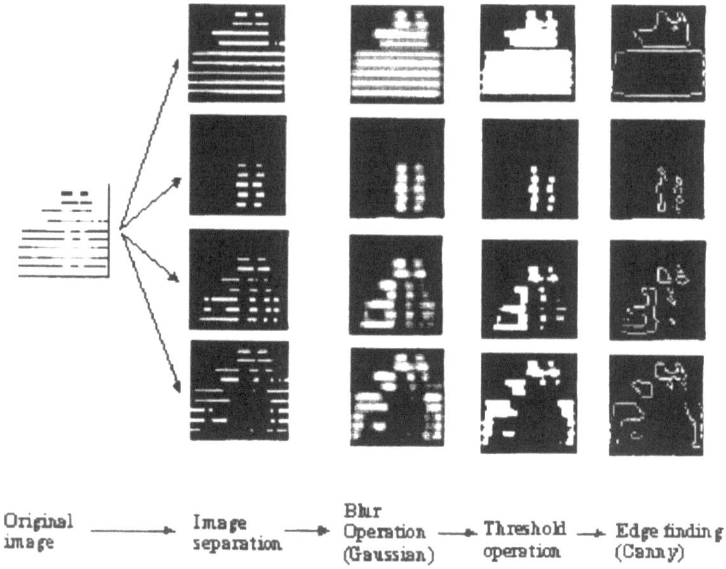

Fig. 5.6 Image processing phases performed on the original ultrasonic images

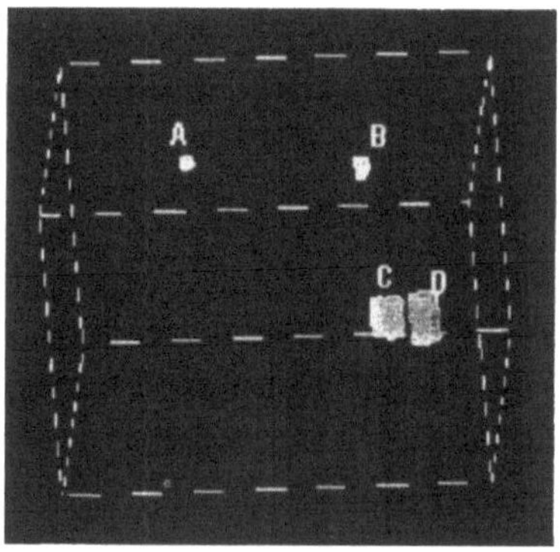

Fig. 5.7 Result of the data fusion operation performed on the C and D-scans from figure 5.2, and reconstructed 3-D defects

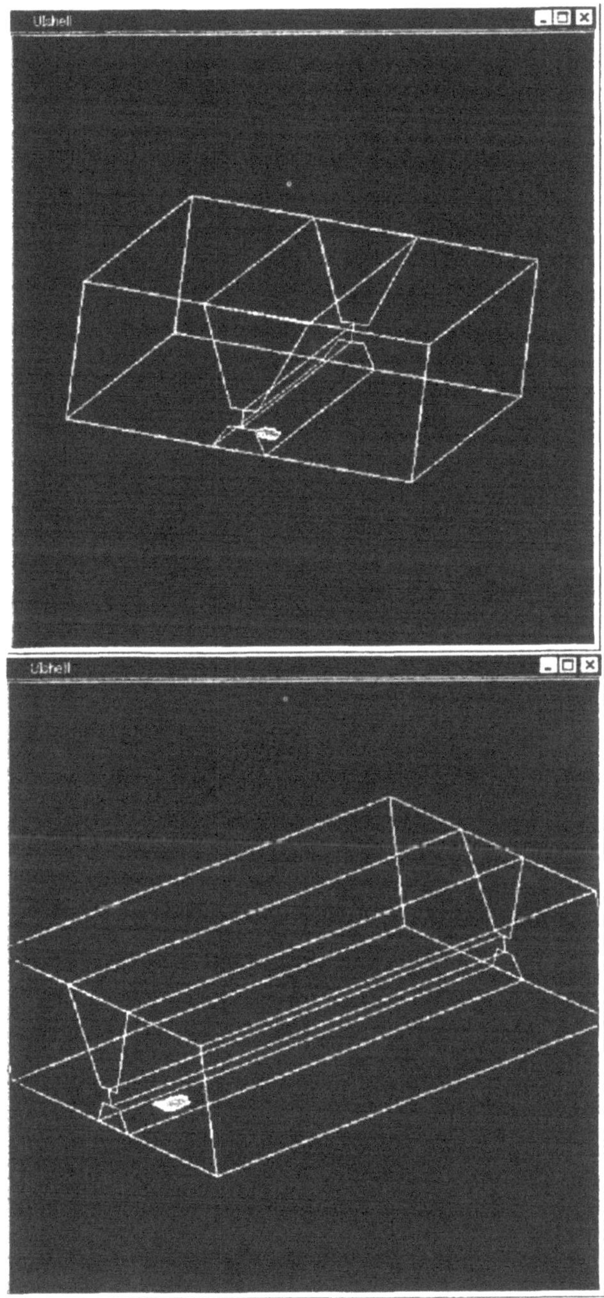

Fig. 5.8 Result of the combination of CAD representation of the material and reconstructed 3-D defect

5.4 CONCLUSION

With increasing use of multisensor systems and more detailed inspection results in digital format, the volume and complexity of data generated during a NDI can become overwhelming. Data fusion is a possible approach for efficient data analysis and data management in NDT. An optimum multisensor data fusion system should be able to perform automatic defect recognition from several ultrasonic views, carry out the fusion process and display a 3-D image of the defect. In addition, it should be able to deal with faulty sensors, operate in the event of one or multiple sensors being non-operational, and perform the combination of information from the remaining active sensors. Based on such expectations, it is thought that the synergistic use of various technologies such as NDT, robotics, data fusion and virtual display, will very certainly become more widespread in the 21st century.

The information presented in this chapter showed that 3-D image reconstruction based on the combination of multiple images, provides a virtual insight into structures and objects, otherwise unavailable. There are great hopes that results of this research work will be put into practice in Hungary in the near future. And that it will benefit the whole nuclear industry by guaranteeing the integrity of welds.

REFERENCES

1. Pons F., Tomasino R., *The French approach to performance demonstration in the context of service inspection of nuclear power plants*, Insight, 1995, 37(3): 159-162.
2. Pitt F.D., Kenny S.D., *Controlling NDT to minimise downtime and risk and to comply with regulatory requirements on 500MW plant*, Insight, 1994, 36(4): 206-209.
3. Rylander L., Gustafsson J., *Non-destructive examination of the primary system in Ignalia nuclear power plant*, Insight, 1994, 36(4): 210-212.
4. Fücsök F., Gubicza P., *Connections of ultrasonic and acoustic emission testing of reactor pressure vessels*, IAEA, Training Course In-Service Inspection Techniques and Systems for VVER-4 Type Reactors at Paks NPP, 1993, Hungary.
5. O'Brien D., *Cutting to the heart of 3-D data*, Scientific Computing World, 1999, (46):43.
6. Wedgwood F.A., *Data processing in ultrasonic NDT*, Proc. of 1987 Ultrasonics International Conf., 1987, London, UK, 381-388.
7. Hirasawa H., Taketomi T., Kuboyama T., Watanabe Y., *Development of a three-dimensional ultrasonic testing system using computer graphics*, Proc. 13th Inter. Conf. on NDE in the Nuclear and Pressure Vessel Industries, 1995, Kyoto, Japan, 153-156.
8. Stolk A.Th., *Reconstructing 3D geometry from an ultrasonic NDT signal*, Insight, 1999, 41(9):596-598.
9. Pastorino M., *Inverse-scattering techniques for image reconstruction*, IEEE Instrumentation & Measurement Magazine, 1998, 1(4):20-25.
10. Capineri L., Windsor C., Zinno F., *Three-dimensional buried mine positions from arbitrary radar surface scans*, Insight, 1999, 41(6):372-375.
11. Summa D.A., Claytor T.N., Jones M.H., Schwab M.J., Hoyt S.C., *3-D visualisation of X-ray and neutron computed tomography (CT) and full waveform ultrasonic (UT) data*, Proc. Review of Progress in QNDE, 1999, Snowbird, USA, 18A:927-934.
12. Suzuki T., Nagai S., Maruyama F., Furukawa H., *High performance acoustic three-dimensional*

image processing system, Proc. 13th Inter. Conf. on NDE in the Nuclear and Pressure Vessel Industries, 1995, Kyoto, Japan, 277-282.
13. Kamgar-Parsi B., Rosenblum L.J., Belcher E.O., *Underwater imaging with a moving acoustic lens*, IEEE Trans. on Image Processing, 1998, 7(1):91-92.
14. Gautier S., Lavayssière B., Fleuet E., Idier J., *Using complementary types of data for 3D flaw imaging*, Proc. Review of Progress in QNDE, 1998, San Diego, USA, 17A:837-843.
15. Schmitz V., Walte F., Chakhlov S.V., *3D ray tracing in austenite materials*, NDT&E Inter., 1999, 32(4):201-213.
16. Koehler B., Hentges G., Mueller W., *Improvement of ultrasonic testing of concrete by combining signal conditioning methods, scanning laser vibrometer and space averaging techniques*, NDT&E Inter., 1998, 31(4):281-287.
17. Prémel D., Madaoui N., Venard O., Savin E., *An eddy current imaging system for reconstruction of three-dimensional flaws*, Proc. 7th Euro. Conf. on NDT, 1998, Copenhagen, Denmark, 3:2584-2591.
18. Vengrinovich V., Denkevich Y., Tillack G.R., *Limited projections and views Bayesian 3D reconstruction using Gibbs priors*, Proc. 7th Euro. Conf. on NDT, 1998, Copenhagen, Denmark, 3:2371-2378.
19. Vengrinovich V.V., Denkevich Y.B., Tillack G.R., Jacobsen C., *3D X-ray reconstruction from strongly incomplete noisy data*, Proc. Review of Progress in QNDE, 1999, Snowbird, USA, 18A:935-942.
20. Vengrinovich V.L., Denkevich Y.B., Tillack G.R., Nockemann C., *Multi step 3D x-ray tomography for a limited number of projections and views*, Proc. Review of Progress in QNDE, 1997, Brunswick, USA, 16A:317-323.
21. Tichonov A.N., Arsenin V.Y., *Solution of ill-posed problems*, Wiley, New York, USA, 1977.
22. Miller K., SIAM J. on Mathematical Analysis, 1970, 1(52).
23. Linden von der W., Appl. Phys. A Materials Science and Processing, *Maximum-entropy data analysis*, 1995, 60(2):155.
24. Saparin G.V., Obyden S.K., Ivannikov P.V., *A nondestructive method for three-dimensional reconstruction of luminescence materials: principles, data acquisition, image processing*, Scanning, 1996, 18: 281-290.
25. Spedding V., *Software combines microscope images*, Scientific Computing World, 1997, (32):27-28.
26. Gros, X.E., Fücsök F., *Combining NDT signals improves safety in nuclear power stations*, Proc. Inter. Conf. In-Service Inspection, 1995, Pula, Croatia.

6 DATA FUSION FOR INSPECTION OF ELECTRONIC COMPONENTS

James REED[1], Seth HUTCHINSON[2]
[1]The University of Michigan,
Ann Arbor, USA
[2]University of Illinois,
Urbana-Champaign, USA

6.1 INTRODUCTION

Parts with circular features, such as holes, are common in the microelectronics industry. For example, holes used for mounting integrated circuits and other electronic components are found on printed circuit boards. If the holes in these circuit boards are not located or shaped properly, electronic components may not fit into them correctly. Due to the small size of many electronic components, holes must be manufactured precisely. Therefore, inspecting the circular shape of these holes requires a high degree of accuracy.

The research described in this chapter is intended for use in automated optical inspection of via holes in printed circuit boards. Via holes are used to provide electrical connections between different sides or layers of a printed circuit board. In such inspection system, circuit boards move along a conveyer at a fixed velocity. Several images are taken of each part as it passes beneath the camera, creating an image sequence that can be defined as: $Q = \{q_1 \ldots q_n\}$. The vias are inspected one-at-a-time, and the entire shape of each inspected hole is visible in every image in Q. Images in the image sequence are perspective projections of the scene; therefore, the via holes appear as ellipses in these images [1]. Given the image sequence, Q, our task is to estimate the parameters of the elliptical shape of the via holes with subpixel accuracy. From these estimates, one can infer the properties of the shape of the actual via holes, and use this information to decide whether a via hole is properly shaped.

The method combines image fusion (including image registration and image enhancement) with subpixel edge detection, and subpixel parameter estimation of ellipses to perform the inspection task described above. In sections 6.2 and 6.3 we describe how, given the input image sequence Q, image enhancement using Peleg and Irani's super-resolution algorithm is performed [2]. The super-resolution algorithm creates a high resolution image H that has twice the resolution of the individual images in Q. In section 6.4 we describe how subpixel arc-edge detection is performed on the high resolution estimate H, yielding a list of data points. The arc-edge detector is a sample-moment-based edge detector that locates data points that lie on a circular arc with subpixel accuracy [3]. The use of these data points by an ellipse parameter estimation algorithm is described in section 6.5 [4]. Among the benefits of this system are increased noise tolerances and reduced hardware requirements. Because image sequence analysis is used for image enhancement, high resolution cameras and high precision positioning equipment are not required. The research was conducted using off-the-shelf hardware and tested on real images.

6.2 IMAGE REGISTRATION

Given a sequence of images $Q = \{q_1 \ldots q_n\}$ our first task is to bring these images into registration, producing a new image sequence, Q_a, in which all images are aligned. In the image sequence Q_a, the pixels corresponding to a circular feature being inspected occupy the same locations in each image. The registration process described in this section was introduced by Irani and Peleg [2].

Image registration is accomplished by estimating the motion vector for each image, then shifting each image according to its motion vector. The motion vectors are estimated in an iterative process and expressed with respect to a reference image, q_r, chosen from Q. For a particular image q_c, let $T = \{t_x + \rho_x, t_y + \rho_y\}$ represent an initial estimate of the motion between images q_r and q_c, where (t_x, t_y) is the integer part of the motion and (ρ_x, ρ_y) is the fractional part. Motion vector estimation is performed by repeating the following steps. First, we shift the image according to its motion estimate. Next, a correction to the motion estimate by solving a system of linear equations given by equation (6.3) is computed. When changes to the motion estimate are less than a threshold value, motion estimation stops.

Because each motion vector has both an integer and a subpixel part, an image is shifted in two steps. In the first part of the shifting operation, q_c is shifted according to the integer part of the motion. No intensity values are changed during this integer shift; therefore, the intensity of each pixel after shifting by (t_x, t_y) can be expressed as:

$$I'(x+t_x,\ y+t_y) = I(x,y) \tag{6.1}$$

where $I(x,y)$ is the intensity of the pixel at position (x,y), and $I'(x+t_x,\ y+t_y)$ is the intensity of the pixel at position ($x + t_x$, $y + t_y$) in the shifted image.

In the second part of the shifting process, the subpixel part of the motion is used to estimate new intensities for the shifted pixels. This is accomplished by approximating the intensity function, $I(x,y)$, by a plane, and using bi-linear interpolation to compute the intensity values at subpixel locations. To avoid the accumulation of errors, the motion estimation algorithm always shifts the original image q_c.

In the second part of the motion estimation algorithm, a correction to the current motion estimate is calculated by solving a set of linear equations derived by Keren *et al.* [5]. In general, motion in two dimensions can be described by a translation (a,b) and a rotation, θ, assumed to be about an axis at the centre of the image. In terms of the motion vector *T*, the components of the motion are $a = t_x + \rho_x$ and $b = t_y + \rho_y$. In terms of the parameters *a*, *b*, θ, the relationship between q_r and q_c is given by:

$$q_c(x,y) = q_r(x\cos\theta - y\sin\theta + a,\ y\cos\theta + x\sin\theta + b) \tag{6.2}$$

As shown in [5], an approximation to the sum-of-squares error between q_r and q_c can be derived by linearisation of (6.2). Setting partial derivatives of this approximate error function to zero, the following system can be obtained:

$$\begin{aligned} \Sigma D_x^2 a + \Sigma D_x D_y b + \Sigma A D_x \theta &= \Sigma D_x D_t \\ \Sigma D_x D_y a + \Sigma D_y^2 b + \Sigma A D_y \theta &= \Sigma D_y D_t \\ \Sigma A D_x a + \Sigma A D_y b + \Sigma A^2 \theta &= \Sigma A D_t \end{aligned} \tag{6.3}$$

where

$$D_x = \frac{\partial q_r(x,y)}{\partial x},$$
$$D_y = \frac{\partial q_r(x,y)}{\partial y},$$
$$D_t = q_c(x,y) - q_r(x,y),$$
$$A = xD_y - yD_x$$

Solving (6.3) for a, b, and θ yields a vector that is used to update the current motion estimate at the end of each iteration, until the magnitude of the correction vector is below a predefined threshold. The motion estimation algorithm produces an aligned sequence of images Q_a, and list of motion vectors L.

6.3 CREATING A HIGH RESOLUTION IMAGE

The super-resolution method of Irani and Peleg [2] was used to create a high resolution image using the aligned image sequence, Q_a, and the list of motion vectors L. The result is a fused image that has twice the resolution of the input images. An alternative approach is given by Jacquemod *et al.* who describes a system that uses subpixel camera displacements to create the high resolution image [6].

The super-resolution algorithm works by creating an initial estimate of the high resolution image, H, and then using an iterative refinement procedure to improve that estimate by exploiting known characteristics of the imaging process. The iterative refinement proceeds as follows. A sequence of low resolution images $S = \{s_1 \ldots s_n\}$ is created by subsampling H, and then shifting each subsampled image according to the corresponding motion vector in L. If the high resolution estimate is correct, then the actual and simulated image sequences will be identical, i.e., $S = Q$. If $S \neq Q$ the difference images between Q and S are calculated, creating a sequence of difference images: $Q - S = \{(q_1 - s_1) \ldots (q_n - s_n)\}$. Corrections to the high resolution estimate are based on the values of these difference images, as described below in section 6.2.

6.3.1 Computing Simulated Low Resolution Images

Let $q(x,y)$ represent a low resolution image pixel. Pixel $q(x,y)$ corresponds to a photosite in the CCD camera [1]. The intensity at $q(x,y)$ is determined by the receptive field of the photosite, which is sensitive to light emanating from a scene region that is defined by the centre, size, and shape of the photosite's receptive field [7]. The receptive fields of adjacent photosites overlap due to the proximity and spacing of the sites; therefore, light emanating from any given points in the scene influences the intensity of several pixels in the low resolution image. The point-spread function of the sensor describes which low resolution pixels are influenced by a given point in the scene. Because CCD cameras use an array of photosites, it is assumed that the receptive fields and point-spread functions of the photosites are identical.

The camera produces a discreetised, low resolution version of the original scene. The imaging model is a mathematical representation of the imaging process.

The model presented by Irani and Peleg [8] was used:

$$q_k(x,y) = \sigma_k(h(f(i,j)) + \eta_k(i,j)) \tag{6.4}$$

where q_k is the kth image of the image stream Q. Equation (6.4) represents a transformation from the scene f, to a digital image q_k. The blurring operator, h, is defined by the point spread function of the sensor. Because we do not actually know the sensor's properties, we assume that the point-spread function is a Gaussian smoothing operator. Additive noise is represented by η_k. The function σ_k digitises the image into pixels and quantifies the resulting pixel intensities into discrete grey levels.

The imaging process model (6.4) describes how low resolution input images are produced. To create simulated low resolution images, equation (6.4) can be approximated by:

$$s^n(x,y) = \sum_{\beta} H^n(i,j)h^{PSF}(i-z_x,\ j-z_y) \tag{6.5}$$

in which s^n is a low resolution image, and H^n is the high resolution image produced in the nth iteration of the refinement algorithm. Each simulated low resolution pixel $s^n(x,y)$ is the weighted average of the high resolution pixels, $H^n(i,j)$, that are in the low resolution pixel's receptive field; the set of these high resolution pixels is denoted by β. The point-spread function of the imaging system, h in equation (6.4), is represented by a mask that is denoted by h^{PSF} in equation (6.5). The image co-ordinates of the centre of this mask, (z_x,z_y), are used to select the mask value for a given high resolution pixel (i,j).

In equation (6.4) a blurred version of the continuous image f is discreetised to produce input low resolution images, while in equation (6.5) a discreetised high resolution image H^n is blurred to produce simulated low resolution images. Each simulated low resolution image s_i^n is created by shifting s^n according to the motion estimate that corresponds to the image q_i in Q. This shifting operation produces a simulated image that is in congruence with q_i. Simulated images are created for each image in the input image stream.

6.3.2 Iterative Refinement

After creating simulated images, improvements to the high resolution estimate are calculated using the updating formula:

$$H^{n+1}(i,j) = H^{n}(i,j) + \left\{ \sum_{k,\alpha} \left(q_k(x,y) - s_k^n(x,y)\right) \frac{\left(h^{BP}(x - z_x', y - z_y')\right)^2}{c \sum_{\alpha} h^{BP}(x - z_x', y - z_y')} \right\} \quad (6.6)$$

where q_k is the kth image of the input image stream Q, and s_k^n is the kth simulated low resolution image. The function h^{BP} represents a back-projection kernel that is used to calculate improvement values for the high resolution estimate. The contribution of the back-projection kernel is normalised using a constant c. The choice of h^{BP} is discussed in the literature [2]. A normalised, 5×5 mask representing a Gaussian operator was used. The proper back-projection kernel value is chosen using the low resolution image co-ordinates of the centre of the back-projection kernel (z_x, z_y). The improvement value for a given high resolution pixel is the weighted average of the contributions of the low resolution pixels in its receptive field; the set of all such pixels is denoted by α. Given the high resolution estimate in the nth iteration of the refinement algorithm H^n, the refinement algorithm creates a new high resolution estimate, H^{n+1}, by calculating improvement values for the pixels of H^n. The initial high resolution estimate H is created by averaging the values of pixels in the aligned image sequence $Q_a = \{q_{a1} \cdots q_{an}\}$:

$$\begin{aligned} H(2x,2y) &= H(2x,2y+1) = H(2x+1,2y) = H(2x+1,2y+1) \\ &= \frac{1}{n} \sum_{q_a \in Q_a} q_a(x,y) \end{aligned} \quad (6.7)$$

6.4 SUBPIXEL EDGE DETECTION

There are many methods of edge detection [9]. Standard edge operators are easy to implement; however, in their simplest forms they are pixel level edge detectors, which can only localise edges to the nearest pixel. Although efforts have been made to increase the accuracy of these methods, they cannot be used for subpixel edge detection unless some form of interpolation is applied [10]. The limited resolution of early edge detectors led to the development of subpixel edge detection algorithms, which localise edges to within a fraction of a pixel precision. A number of subpixel edge detectors rely on some form of interpolation [11-13], while others rely on moment analysis [3,8,10,14,15]. The arc-edge detector described by Tchoukanov *et al.* was used [3]. In the first step, bilevel thresholding and simple edge detection were applied to the high resolution image H to create an edge map. The following operations are performed on H at each location specified in the edge map. First, the circular arc with straight-line segments is approximated. The parameters of these line segments are calculated to subpixel accuracy using Tabatabai and Mitchell's moment-based straight-line-edge detector [10]. Given the straight-line approximation of the circular shape, the co-ordinates of the points that lie on the circular border curve are calculated. These data points are used in the ellipse parameter estimation described in section 6.5.

6.4.1 Building the Initial Edge Map

The arc-edge detector is applied to points of interest in the high resolution image *H*. These points are indicated on an edge map that is created by performing bilevel thresholding on *H*, then using simple edge detection to locate edge points in the binary image. Tsai's sample moment preserving bilevel thresholding algorithm was used to threshold *H* [16]. Given an input image, this algorithm locates the threshold value t such that the first four sample moments of the image histogram are preserved. The ith sample moment of the image data is:

$$M_i = \frac{1}{n}\sum_x \sum_y \left(H(x,y)\right)^2 \qquad (6.8)$$

where *H*(x,y) is the pixel intensity at location (x,y) in *H*, and n denotes the number of pixels in *H*. By this definition, $M_0 = 1$. Sample moments can also be computed from the histogram of *H*:

$$M_i = \frac{1}{n}\sum_j n_j z_j^i = \sum_j p_j z_j^i \qquad (6.9)$$

where n is the number of pixels in the image, n_j is the number of pixels in *H* with intensity value z_j, and $p_j = \frac{n_j}{n}$. For bilevel thresholding, let z_b and z_f be the representative intensity values for the two regions of a binary image. The moment-preserving thresholding algorithm selects a threshold such that if the below threshold pixels are replaced by z_b and the above threshold pixels are replaced by z_f, then the first four sample moments will be preserved in the resulting bilevel image. Let u_b represent the fractions of below threshold pixels in *H*, and u_f represent the fractions of above threshold pixels in *H*, with $u_b + u_f = 1$. The first four sample moments of the bilevel image are given by:

$$M_i' = (z_b)^i u_b + (z_f)^i u_f \ , \ i = 0, 1, 2, 3 \qquad (6.10)$$

To preserve the first four moments in the bilevel image, $M_i' = M_i$ for $i = 0, 1, 2, 3$, leading to:

$$
\begin{aligned}
&u_b + u_f = 1 \\
&z_b u_b + z_f u_f = \sum_j p_j z_j \\
&(z_b)^2 u_b + (z_f)^2 u_f = \sum_j p_j z_j^2 \\
&(z_b)^3 u_b + (z_f)^3 u_f = \sum_j p_j z_j^3
\end{aligned}
\tag{6.11}
$$

The system (6.11) is solved for z_b, z_f, u_b and u_f. The threshold value t is chosen by examining the histogram of the image. Beginning with the lowest intensity value in the histogram, we begin summing the number of pixels of each intensity value. The threshold value is chosen as the first intensity value that makes this sum of pixels greater than or equal to the fraction of below threshold pixels u_b. Given t, thresholding on H is performed to produce a binary image. Simple edge detection is performed on the binary image to produce the edge map.

6.4.2 Straight-Line Edge Detection

Given the high resolution estimate H and the edge map, sample moment preserving straight-line-edge detection is performed on H at locations specified by the edge map using Tabatabai and Mitchell's subpixel straight-line-edge detector [3,10]. The procedure is summarised next.

The straight-line-edge detector locates lines that lie within a circular detection area. The detection circle consists of 69 pixels that are weighted to approximate a circle of radius 4.5 pixel units. Each pixel in the detection circle is weighted by the amount of the circle area it occupies. For each location (x,y) stored in the edge map, the detection area is centred at location (x,y) and perform straight-line-edge detection in H.

The parameters of the straight-line-edge model are shown in figure 6.1. Let r represent the radius of the detection circle. A straight-line-edge divides the detection circle into the two regions A_1 and A_2. We assume that the edge can be modelled as a step, with pixels of a given intensity on one side of the edge and pixels of a different intensity on the other. These intensities are the characteristic intensities of the regions that border the straight-line-edge.

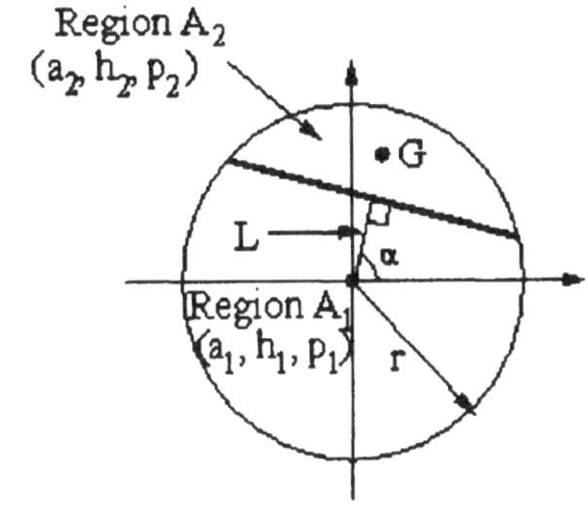

Fig. 6.1 Parameters of the edge model

The two regions that border the edge are described by a_i, the area of the region i; h_i, the characteristic intensity of region i; and p_i, the relative frequency of occurrence of pixels with intensity h_i within the detection circle. The straight-line-edge detector locates the edge by solving for the unknowns a_i, h_i, and p_i.

To facilitate this discussion, consider a line segment drawn from the centre of the detection circle to the straight-line-edge, and normal to the straight-line-edge (figure 6.1). The length of this normal, denoted by L, is the perpendicular distance from the centre of the detection circle to the straight-line-edge. The angle of orientation of the straight-line-edge, α, is defined as the angle between the normal and the horizontal axis of the detection circle. In the following discussion, we will refer to the centre of gravity G of A_2.

The straight-line-edge detector is also based on sample moments. Edges are located such that the first four sample moments of the image data are preserved. This is similar to the moment-preserving thresholding that was used in creating the edge map. In creating the edge map, the moments were calculated for the entire image H. Here, the moments are calculated for the pixels within the detection circle. The first four sample moments are defined as:

$$M_i = \sum_x \sum_y (H(x,y))^i w(x-d_x, y-d_y)$$
$$i = 0, 1, 2, 3; \quad (x,y) \in \text{detection circle} \tag{6.12}$$

The weight of pixel $H(x,y)$ is denoted by $\omega(x - d_x, y - d_y)$, where d_x, d_y is the centre of the detection circle. The summation (6.12) is taken over the pixels in the detection circle. The equations that describe the sample moments used in edge detection are similar to the equations used for bilevel thresholding in section

6.4.1. To preserve the first four sample moments, the following relations must be satisfied:

$$\begin{aligned} p_1 + p_2 &= M_0 \\ h_1 p_1 + h_2 p_2 &= M_1 \\ (h_1)^2 p_1 + (h_2)^2 p_2 &= M_2 \\ (h_1)^3 p_1 + (h_2)^3 p_2 &= M_3 \end{aligned} \tag{6.13}$$

where h_i, and p_i are described above. Tabatabai and Mitchell present the following solutions for equation (6.13) in [10]. After computing the first four moments, the frequencies of occurrence p_1 and p_2 are calculated:

$$p_1 = 1 - p_2 \tag{6.14}$$

$$p_2 = \frac{1}{2}\left[1 + \bar{s}\sqrt{\frac{1}{4 + \bar{s}}}\right] \tag{6.15}$$

where:

$$\bar{s} = \frac{M_3 + 2M_1^3 + 2M_1 M_2}{\bar{\sigma}^3} \tag{6.16}$$

$$\bar{\sigma} = M_2 - M_1^2 \tag{6.17}$$

The characteristic intensities are calculated using p_1 and p_2 :

$$h_1 = M_1 - \bar{\sigma}\sqrt{\frac{p_2}{p_1}} \tag{6.18}$$

$$h_2 = M_1 + \bar{\sigma}\sqrt{\frac{p_1}{p_2}} \tag{6.19}$$

After computing p_1, p_2, h_1 and h_2 for the pixels within the detection circle, we determine whether an edge is present by using a threshold on the difference between the characteristic intensities of A_1 and A_2. A lower bound for this threshold is derived using equation (6.18) and (6.19):

$$\frac{(h_1 - h_2)^2}{\bar{\sigma}^2} = \frac{1}{p_1 p_2} \tag{6.20}$$

Because $p_1 + p_2 = 1$, and $p_1 p_2 \leq \frac{1}{4}$:

$$(h_1 - h_2)^2 \geq 4\bar{\sigma}^2 \tag{6.21}$$

which yields:

$$|h_1 - h_2| \geq 2\bar{\sigma}^2 \tag{6.22}$$

If equation (6.22) is satisfied, it can be said that there is an edge within the detection circle. Once an edge has been detected, its angle of orientation, α, and normal distance from the centre of the detection circle, *L*, are calculated (figure 6.1). Edges are described by their normal equation with respect to the co-ordinate system of the detection circle:

$$\begin{aligned} x\cos\alpha + y\sin\alpha &= -L \text{ if } p_1 \leq p_2 \\ x\cos\alpha + y\sin\alpha &= L \text{ otherwise} \end{aligned} \tag{6.23}$$

Values for cosα and sinα are calculated using the centre of gravity of the data within the detection circle. The co-ordinates of the centre of gravity are calculated using:

$$G_x = \frac{1}{M_1} \sum_x \sum_y x H(x,y) \omega(x - d_x, y - d_y) \tag{6.24}$$

$$G_y = \frac{1}{M_1} \sum_x \sum_y y H(x,y) \omega(x - d_x, y - d_y) \tag{6.25}$$

The summations in equations (6.24) and (6.25) are performed for all pixels (x,y) within the detection circle. The angle α can be calculated using:

$$\tan\alpha = \frac{G_y}{G_x} \tag{6.26}$$

Referring to figure 6.1, the normal distance *L* is found by calculating the area enclosed between the edge line and the detection circle, i.e. the area a_2 of region A_2. As shown by Tchoukanov *et al.* [3] this leads to the following solutions for *L*:

$$r^2 \arcsin\left(\frac{\sqrt{r^2 - L^2}}{r}\right) - L\sqrt{r^2 - L^2} - a_2 = 0 \text{ if } p_1 \geq p_2 \quad (6.27)$$

$$0.5\pi r^2 - L\sqrt{r^2 - L^2} - r^2 \arcsin\left(\frac{L}{r}\right) - a_2 = 0 \text{ otherwise} \quad (6.28)$$

6.4.3 Arc-edge Data Point Calculation

In the previous section the circular border curve was approximated using line segments defined by the parameters α and *L*. In this section, we will show how these parameters are used to calculate the co-ordinates of the data points that lie on the circular border curve. These data points will be called arc-edge data points. As shown in figure 6.2, arc-edge data points are the intersection points of the straight-line approximation of the border curve and the border curve itself.

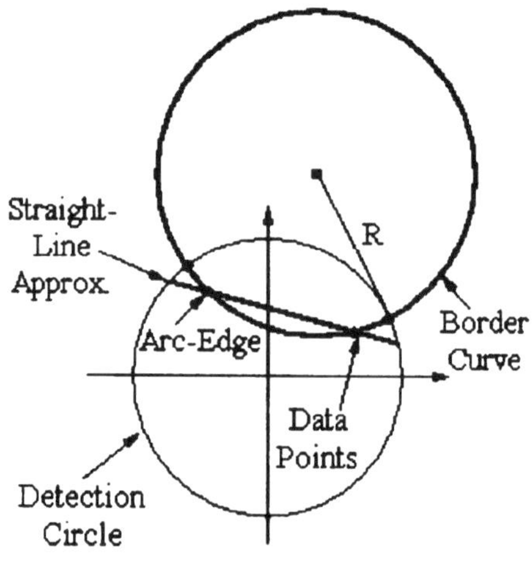

Fig. 6.2 Arc-edge data points

It is assumed that the border curve can be approximated by a circle of radius *R*, which can be described by:

$$(x - X_0)^2 + (y - Y_0)^2 = R^2 \quad (6.29)$$

where (X_0, Y_0) are the co-ordinates of the centre of the circle. Tchoukanov *et al.* derived a geometric relationship between the locations of the arc-edge data points and the parameters of the approximating line, based on the assumption that the position of the arc-edge data points is a weak function of *R* (i.e. the choice of *R* has so little effect on the position of the arc-edge data points that it can be ignored) [3]. This assumption is used to derive equations that allow us to calculate the co-ordinates of the arc-edge data points. The following derivation of these equations follows that of Tchoukanov *et al.* [3].

Figure 6.3 illustrates the arc-edge data points in greater details. The circular border curve is centred at the point (X_0, Y_0), has radius *R*, and intersects the detection circle at the points (x_1, y_1) and (x_2, y_2). The arc-edge data points are located at (x_3, y_3) and (x_4, y_4). Equations are then derived to calculate the co-ordinates of (x_3, y_3) and (x_4, y_4), given *L* and α.

To facilitate this derivation, the detection circle is rotated counter-clockwise through an angle of $(0.5\pi - \alpha)$ to align the normal *L* with the Y-axis. Let (X_0', Y_0') be the co-ordinates of the centre of the rotated circular arc. The co-ordinates of the rotated arc-edge points, (x_3', y_3') and $(-x_3', y_3')$, are given by:

$$x_3' = x_3 \sin\alpha - y_3 \cos\alpha \tag{6.30}$$

$$y_3' = L = x_3 \cos\alpha + y_3 \sin\alpha \tag{6.31}$$

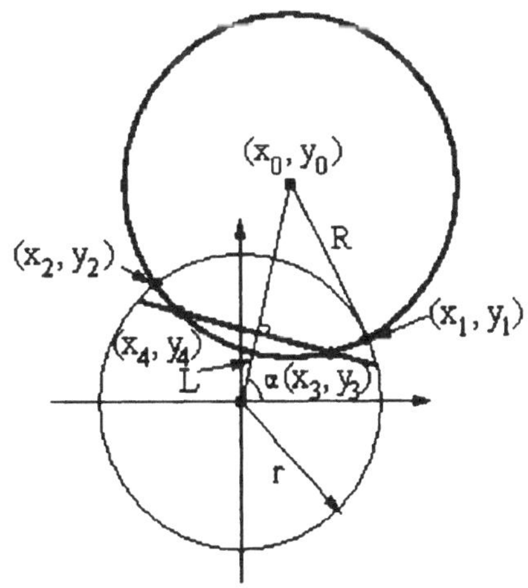

Fig. 6.3 Parameters used in arc-edge data points calculation

The area of region A_2 is:

$$a_2 = p_2 r^2 \pi = 2 \int_0^{x_1'} \int_{Y_0' - \sqrt{R^2 - x^2}}^{\sqrt{r^2 - L^2}} dx dy \tag{6.32}$$

Tchoukanov *et al.* derived the following equations for x_3' by solving the integral in equation (6.32):

$$\begin{aligned} & r^2 \arcsin\frac{x_1'}{r} + R^2 \arcsin\frac{x_1'}{r} - x_1' Y_0' - a_2 = 0 \quad \text{if } p_1 \geq p_2 \\ & 0.5\pi r^2 + r^2 \arccos\frac{x_1'}{r} + R^2 \arcsin\frac{x_1'}{r} - x_1' Y_0' - a_2 = 0 \quad \text{otherwise} \end{aligned} \tag{6.33}$$

where:

$$Y_0' = L + \sqrt{R^2 - x_3'^2} \tag{6.34}$$

$$x_1' = \sqrt{r^2 - \left(\frac{r^2 - L^2 + 2LY_0' - x_3'^2}{2Y_0'} \right)^2} \tag{6.35}$$

Tchoukanov *et al.* [3] have used the approximation:

$$K = \frac{x_3'}{\sqrt{r^2 - L^2}} \tag{6.36}$$

where a lookup table of *K* values is created off-line, indexed on values of *L* in the range (-4.5 … +4.5) pixel units. The co-ordinates of the arc-edge data points (x_3, y_3) and (x_4, y_4), are calculated with:

$$x_3 = L\cos\alpha + K\sqrt{r^2 - L^2}\sin\alpha \tag{6.37}$$

$$y_3 = L\sin\alpha - K\sqrt{r^2 - L^2}\cos\alpha \tag{6.38}$$

$$x_4 = L\cos\alpha - K\sqrt{r^2 - L^2}\sin\alpha \tag{6.39}$$

$$y_4 = L\sin\alpha + K\sqrt{r^2 - L^2}\cos\alpha \tag{6.40}$$

The complete subpixel arc-edge detector is illustrated in figure 6.4. Given an input image *H*, bilevel thresholding and simple edge detection are performed to create an edge map. Moment-preserving straight-line-edge detection is performed on *H* at locations specified in the edge map, producing a list of straight-line segments that approximate the circular border curve. Each line segment is defined by its normal distance from the centre of the detection circle *L* and its angle of orientation α. The co-ordinates of the arc-edge data points are calculated with equation (6.37) to (6.40).

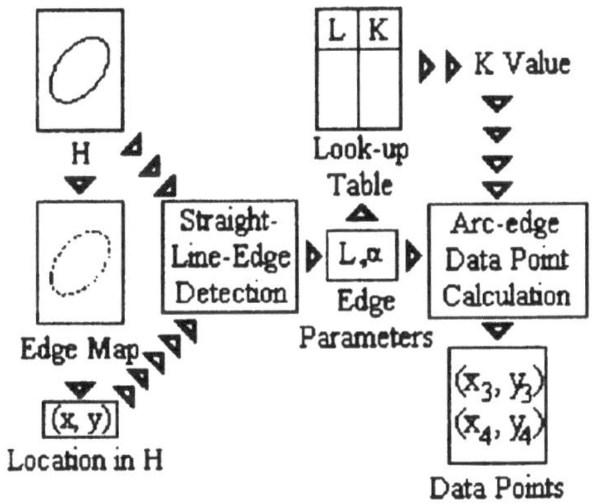

Fig. 6.4 Arc-edge data points calculation

6.5 ELLIPSE PARAMETER ESTIMATION

Under perspective projection, a circle in the scene will appear as an ellipse in the digital image; therefore, the list of data points described in section 6.4 to perform ellipse parameter estimation is used. Let $P = \{P_1 \dots P_n\}$ represents the list of n data points. Given *P*, our task is to estimate the centre point co-ordinates (X_0, Y_0), the major axis length *A*, minor axis length β, and the angle or orientation Θ of the ellipse that fits the data points.

Various methods have been reported for ellipse parameter estimation [4,17-19]. The area-based parameter estimation algorithm described by Safaee-Rad *et al.* was used [4]. Parameter estimation proceeds as follows. In the first step, the parameters of an initial optimal ellipse are estimated. These parameters are used to generate weights for the data points. The weights normalise the contribution of each data point to the parameter estimation. In the final step, the weighted data points are used to find the parameters of the ellipse. The implicit equation of an ellipse can be written as follows:

$$W(X,Y)=0 \tag{6.41}$$

where:

$$W(X,Y)=aX^2+bXY+cY^2+dX+eY+1 \tag{6.42}$$

The typical approach to fit an ellipse to data points is to minimise an error residual $\Im_0$, given by:

$$\Im_0=\sum_{i=1}^{n}[W(X_i,Y_i)]^2 \tag{6.43}$$

It is well known that equation (6.42) does not give the geometric distance from the point (X,Y) to the ellipse given by equation (6.41), and therefore, that minimising equation (6.43) gives an ellipse that minimises the algebraic distance from the ellipse to the data points, not the geometric distance. To correct for this, Safaee-Rad *et al.* introduced an error function that weights the contribution of each data point in the minimisation according to the ratio between the point's geometric and algebraic distance to the ellipse [4]. Their derivation is summarised next.

First, a new error function, $\Im_1$, based on the difference in the areas of two concentric ellipses with equal eccentricity is derived. Next, it is showed that minimising $\Im_1$ is equivalent to minimising (6.43). Finally, we construct $\Im_1'$ by weighting the data points based on the ratio of geometric *vs.* algebraic distance.

Let (A, B, θ, X_0, Y_0) be the parameters of the ellipse that best fits the data points, and (A', B', θ, X_0, Y_0) be the parameters of the ellipse passing through a given data point $P_i = (X_i, Y_i)$. These ellipses will be referred to as the optimal ellipse and the data point ellipse. The two ellipses are concentric and have the same eccentricity and orientation (figure 6.5).

If *D'* is the area of the data point ellipse and *D* is the area of the optimal ellipse, then an error function can be defined as the difference between these areas as:

$$e_i=D-D' \tag{6.44}$$

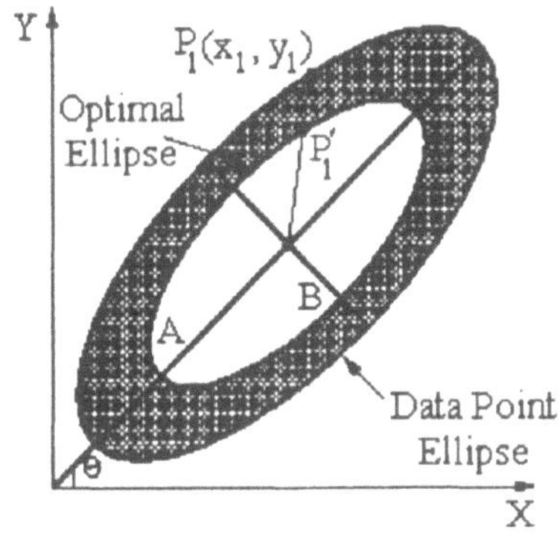

Fig. 6.5 Area difference between two concentric ellipses

Consider a line that passes through a data point $P_i = (X_i, Y_i)$ and the centre point (X_0, Y_0). Let $P_i' = (X_i', Y_i')$ be the intersection point of this line and the optimal ellipse. To aid in this discussion, the following quantities are defined. Let d_i' be the distance from the centre of the ellipse to the point P_i' and let d_i be the distance from the centre of the ellipse to the point P_i.

Given that the two concentric ellipses are similar, i.e. they have the same orientation angle and eccentricity, an expression for the area difference of the ellipses can be derived from equation (6.44) as follows:

$$\begin{aligned}\Delta\text{Area} = e_i &= D - D' \\ &= \pi AB - \pi A'B' \\ &= \pi A'B'\left(\frac{AB}{A'B'}\right)\left(1 - \frac{A'B'}{AB}\right) \\ &= \pi A'B'\left(\frac{d_i^2}{d_i'^2}\right)\left(1 - \frac{d_i'^2}{d_i^2}\right)\end{aligned} \tag{6.45}$$

Bookstein has shown that the following proportionality holds [20]:

$$W(X_i, Y_i) \propto \left[\frac{\delta_i}{d_i'}\left(\frac{\delta_i}{d_i'} + 2\right)\right] \tag{6.46}$$

where, $\delta_i = d_i - d_i^{'}$. After a bit of algebraic manipulation, the following can be written:

$$\left[\frac{\delta_i}{d_i^{'}}\left(\frac{\delta_i}{d_i^{'}}+2\right)\right] = \frac{d_i^2}{d_i^{'2}}\left(1-\frac{d_i^{'2}}{d_i^2}\right) \tag{6.47}$$

Therefore, equation (6.45) is proportional to equation (6.42), and consequently, a new error function, $\Im_1$, can be defined in terms of equation (6.45):

$$\Im_1 = \sum_{i=1}^{n}\left[\frac{1}{\pi A^{'}B^{'}}(D - D_i^{'})\right]^2 = \sum_{i=1}^{n}\left[\frac{e_i}{\pi A^{'}B^{'}}\right]^2 \tag{6.48}$$

Thus, minimising equation (6.48) is equivalent to minimising equation (6.43). Distance $d_i^{'}$ is maximum for points along the ellipse's major axis and minimum for points along its minor axis; therefore, equation (6.45) is maximum for data points near the minor axis of the ellipse and minimum for data points near the major axis of the ellipse. For this reason, the contributions of the data points are normalised by defining a weighting factor that is a function of each data point's position relative to the major axis of the optimal ellipse. Let d_i be the distance of a data point P_i to the optimal ellipse, i.e. distance $\left|\overline{P_i P_i^{'}}\right|$ in figure 6.6.

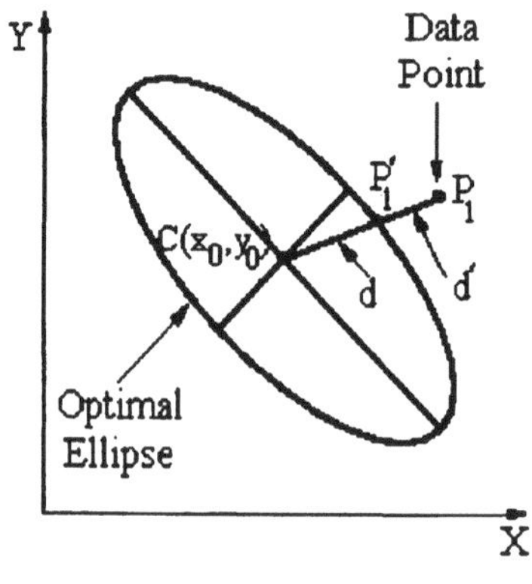

Fig. 6.6 Distances d_I and $d_i^{'}$ of an ellipse

The geometric distance from the centre of the optimal ellipse to the optimal ellipse, distance $\overline{(X_0, Y_0)P'}$ in figure 6.6, is represented by d_i'. Using the expression for the error e_i of a data point given by (6.45), after some algebraic manipulation, we obtain:

$$e_i = \pi A'B'\left(\frac{d_i^2}{d_i'^2}\right)\left(1 - \frac{d_i'^2}{d_i^2}\right) = \pi A'B'\left(\frac{d_i^2}{d_i'^2}\right)\left(\frac{\delta^2 + 2\delta d_i'}{d_i^2}\right) \tag{6.49}$$

Equation (6.49) is the general expression for the error due to data point P_i. This error will be a minimum when P_i is near the major axis of the ellipse. If P_i is on the major axis of the optimal ellipse and has the same distance d_i, the error will be given by e_2:

$$e_2 = \pi A'B'\left(\frac{A^2}{A'^2}\right)\left[\frac{\delta^2 + 2\delta A'}{A^2}\right] \tag{6.50}$$

The ratio e_2/e_i is given by:

$$\frac{e_2}{e_1} = \frac{A^2}{A'^2}\frac{d_i'^2}{d_i^2}\frac{\left[\dfrac{\delta^2 + 2\delta A'}{A^2}\right]}{\left[\dfrac{\delta^2 + 2\delta d_i'}{d_i^2}\right]} = \frac{d_i'}{A'}\frac{\left(\dfrac{\delta}{2A'} + 1\right)}{\left(\dfrac{\delta}{2d_i'} + 1\right)} \approx \frac{d_i'}{A'} = \psi_i \tag{6.51}$$

where the final approximation follows because typically δ is much smaller than either A' or d_i'. We now construct an error function using these weights as follows:

$$\Im_1' = \sum_{i=1}^{n}\left[\psi_i\left(\frac{1}{\pi AB}\right)(D - D_i')\right]^2 = \sum_{i=1}^{n}\left[\psi_i W(X_i, Y_i)\right]^2 \tag{6.52}$$

The error function described in equation (6.52) minimises the area-based error function described by equation (6.44), and normalises the contributions of the data points. Equations for the ellipse parameters can be derived by computing the derivatives of $\Im_1'$ with respect to the unknowns, leading to:

$$X_0 = \frac{2cd - be}{b^2 - 4ac} \tag{6.53}$$

$$Y_0 = \frac{2ae - bd}{b^2 - 4ac} \tag{6.54}$$

$$\theta = \arctan\left[\frac{(c-a) + \sqrt{(c-a)^2 + b^2}}{b}\right] \tag{6.55}$$

$$A^2 = \left[\frac{2(1-F_S)}{b^2 - 4ac}\right]\left[(c+a) + \sqrt{(c-a)^2 + b^2}\right] \tag{6.56}$$

$$B^2 = \left[\frac{2(1-F_S)}{b^2 - 4ac}\right]\left[(c+a) - \sqrt{(c-a)^2 + b^2}\right] \tag{6.57}$$

where:

$$F_S = \frac{bde - ae^2 - cd^2}{b^2 - 4ac} \tag{6.58}$$

Thus, we have a two-stage algorithm. In the first stage, $\Im_1$ is used to generate the parameters of an initial estimate of the ellipse. These parameters are then used to estimate the value of A' and d_1' for each data point. The data point weighting values ψ_i are calculated using A' and d_1'. The weighted data points are used in $\Im_1'$, in the second stage, to find the parameters of the final optimal ellipse. This method of parameter estimation is non-iterative and produces good results.

6.6 EXPERIMENTAL RESULTS

Our method of ellipse parameter estimation was tested on real image sequences of circles with diameters of one-eighth inch, one-fourth inch, one-half inch, one inch, and two inches. The high resolution parameter estimation method produced good results for each circle. Figure 6.7 shows the results of parameter estimation for a one-inch diameter circle.

For each image sequence, our algorithms were used to estimate the major and minor axis lengths for the image ellipses. Given these estimates, in pixel units, an estimate of the axis length in inches is calculated in the following manner. First, the endpoints of each axis are located in the image. Then, Tsai's camera calibration [7] method was used to compute the world co-ordinates that correspond to the axis endpoints. Finally, each axis length is calculated as half the Euclidean distance between the world co-ordinates of its endpoints.

Fig. 6.7 Estimated ellipse plotted onto 1-inch diameter circle

For comparison, results for five different methods of edge detection were obtained. In simple edge detection, moment-based binary thresholding is performed [16], on the high resolution, image produced by the super-resolution algorithm. In the binary image, pixels that border regions of different intensities are treated as data points. Canny 1 edge detection is the Canny edge detector without subpixel interpolation. Canny 2 edge detection is the Canny edge detector with subpixel interpolation. The interpolation is accurate to within a tenth of a pixel dimension. Both the Canny 1 and Canny 2 edge detectors were run on low resolution input images. To investigate the benefit of performing super-resolution, the arc-edge detector was also run on low resolution input images. High resolution edge detection is the method of edge detection that was described in section 6.4 and is used in our method of ellipse parameter estimation. This edge detection method performs arc-edge data point detection on the high resolution image produced by the super-resolution algorithm. In each experiment, parameter estimation was performed using the area-based algorithm described in section 6.5.

Table 6.1 lists the average percent error of the axis length estimates. The percent error, Δ, is calculated using:

$$\Delta = \frac{\left(\left|\text{Estimated A} - \text{R}\right| + \left|\text{Estimated B} - \text{R}\right|\right)}{2\text{R}} \times 100 \qquad (6.59)$$

where R is the actual radius of the circle.

Table 6.1 Average percent error of circle radius estimates

Method	Circle Diameter (inches)				
	0.125	0.25	0.5	1	2
Simple	23.4313	10.3163	5.76656	2.6687	1.51665
Canny 1	22.3480	11.0073	5.92641	2.97288	1.44266
Canny 2	20.9358	10.5905	5.87558	2.91241	1.42978
Arc Edge	22.4000	11.5330	5.14937	2.51762	1.60457
High Res.	1.5660	0.8528	1.04134	1.75634	1.48155

The high resolution method consistently provides accurate parameter estimates. The greatest benefit is seen when inspecting small circles. The high resolution method benefits from both subpixel accuracy and super-resolution. The improvement gained by using the high resolution inspection method decreases as the radius of the circle increases. For large circles, the methods provide virtually the same results.

6.7 CONCLUSION

In this chapter, a new method of parameter estimation for circular shapes that use image sequences was presented. In this method, a sequence of images is used to create a fused high resolution image. A moment-based edge detector that locates with subpixel accuracy points that lie along a circular arc is used to locate data points in the high resolution image, creating a list of data points. Given the list of data points, parameter estimation is performed using an area-based ellipse parameter estimation algorithm. Once the ellipse parameters have been estimated, camera calibration techniques are used to translate distances in the image plane into distances in the real world.

REFERENCES

1. Fu K., Gonzalez R., Lee C., *Robotics: Control, Sensing, Vision, and Intelligence*, New York, McGraw-Hill Pub., 1987.
2. Irani M., Peleg S., *Improving resolution by image registration*, CVGIP Graphical Models and Image Processing, 1991, 53(3):231-239.
3. Tchoukanov I., Safaee-Rad R., Benhabib B., Smith K., *Sub-pixel edge detection for accurate estimation of elliptical shape parameters*, Proc. CSME Mechanical Eng. Forum 1990, Toronto, Canada, 1990, 313-318.
4. Safaee-Rad R., Tchoukanov I., Benhabib B., Smith K., *Accurate parameter estimation of quadratic curves from grey-level images*, CVGIP: Image Understanding, 1991, 54(2):259-274.
5. Keren D., Peleg S., Brada R., *Image sequence enhancement using sub-pixel displacements*, Proc. IEEE Conf. on Computer Vision and Pattern Recognition, 1988, Santa Barbara, USA, 742-746.

6. Jacquemod G., Odet C., Goutte R., *Image resolution enhancement using subpixel camera displacement*, Signal Processing, 1992, 26(1):139-146.
7. Tsai R.Y., *A versatile camera calibration technique for high-accuracy 3D machine vision metrology using off-the-shelf television cameras and lenses*, IEEE Trans. on Robotics and Automation, 1987, RA-3(4):323-344.
8. Irani M., Peleg S., *Image sequence enhancement using multiple motions analysis*, Proc. IEEE Computer Vision and Pattern Recognition, 1992, Champaign, USA, 216-222.
9. Haralick R.M., Shapiro L.G., *Computer and Robot Vision*, Addison-Wesley Pub., 1993.
10. Tabatabai A., Mitchell O.R., *Edge location to subpixel values in digital imagery*, IEEE Trans. on Pattern Analysis and Machine Intelligence, 1984, 6(2):188-201.
11. Petkovic D., Niblack W., Flickner M., *Projection-based high accuracy measurement of straight line edges*, Machine Vision and Applications, 1988, 1(4):183-199.
12. Niblack W., Petkovic D., *On improving the accuracy of the Hough transform*, Machine Vision and Applications, 1990, 3(2):87-106.
13. Carbone J., *Sub-pixel interpolation with charge injection devices*, Proc. SPIE Optical Sensors and Electronic Photography, 1989, Los Angeles, USA, 1071:80-89.
14. Lyvers E.P., Mitchell O.R., Akey M.L., Reeves A.P., *Subpixel measurements using a moment-based edge operator*, IEEE Trans. on Pattern Analysis and Machine Intelligence, 1989, 11(12):1293-1309.
15. Ghosal S., Mehrotra R., *Orthogonal moment operators for subpixel edge detection*, Pattern Recognition, 1993, 26(2):295-306.
16. Tsai W., *Moment-preserving thresholding: a new approach*, Computer Vision, Graphics, and Image Processing, 1978, 29(3):377-393.
17. Tsuji S., Matsumoto F., *Detection of ellipses by a modified Hough transformation*, IEEE Trans. on Computers, 1978, 27(8):777-781.
18. Takiyama R., Ono N., *An iterative procedure to fit an ellipse to a set of points*, IEICE Trans. on Communications Electronics Information and Systems, 1991, 74:3041-3045.
19. Nagata T., Tamura H., Ishibashi K., *Detection of an ellipse by use of a recursive least-squares estimator*, J. of Robotic Systems, 1985, 2(2):163-177.
20. Bookstein F.L., *Fitting conic sections to scattered data*, Computer Graphics and Image Processing, 1979, 9(1):56-71.

7 NDT DATA FUSION IN THE AEROSPACE INDUSTRY

James M. NELSON, Richard H. BOSSI
The Boeing Company,
Seattle, USA

7.1 NDT DATA FUSION IN THE AEROSPACE INDUSTRY

The process of integrating data from diverse process measurements and non-destructive observations of a known object into a consistent description of the condition of the object is an important function of aerospace manufacturing review board and failure investigations. NDE data fusion methods commonly used in aerospace applications include procedures and software to co-register, collocate, and combine multiple NDT data sets into useful, condensed forms, often as a key input to critical contracting, delivery, deployment, and or production decisions. While co-registration and collocation of data at the instrument or sensor level is a potentially useful implementation of NDT data fusion processes, it is often impractical for these types of aerospace decision processes. This is because the primary aerospace applications for NDT data fusion typically involve historical data, remote process steps, the use of extremely large and complex testing facilities, and both national security and proprietary issues. Consequently, the emphasis of NDT data fusion method development has been toward development of post-process NDT data fusion systems and methods where workstation based co-registration and collocation of diverse sets of NDT, engineering, and process data sets is the key technical functionality.

Despite the focus of aerospace NDT data fusion on post-process methods, other forms of sensor-based data fusion are commonly employed in aerospace for mixing of image and/or sensor data from scene or target recognition systems on operational platforms. The aerospace community is deeply committed to research and development of efficient, real time sensor based data fusion methods, primarily aimed at operational applications such as military target recognition. Airborne or orbital surveillance platforms such as missiles or military satellites need image data fusion for data compression and/or faster target recognition. In these cases, the sensor and/or imaging systems are well defined and highly integrated with the fusion algorithms. In these cases, the object of data fusion algorithms is to reduce the processing time or storage requirement associated with the extraction of information contained in the data.

An example of image mixing based data fusion is shown in figure 7.1. In this example, wavelet transform methods have been used to combine a forward looking infrared (FLIR) image and a video image of a F-15 aircraft taking off from St. Louis' Lambert Field. The combined image contains important features of both independent images. By combining images of this type viewed from a common reference frame, the workload on a scene recognition processing system can be reduced, and consequently the time for making critical mission decisions may be reduced.

Fig. 7.1 Mixing of video and FLIR images of a F-15 aircraft taking off

Image compression and fast object (defect) recognition are also desirable features of NDE data fusion algorithms for aerospace applications. However, post-process NDE data fusion algorithms must provide unique additional functionality. NDE data fusion must take into account the geometry of both the inspection systems as well as the part being inspected to achieve correct registration of data. While camera based or common scanner based multimode inspections are effective ways to achieve convenient multimode data registration, they rarely are implemented for aerospace NDT applications. Hand-held and camera-based NDT systems have most applicability to the evaluation of ageing aircraft, particularly for corrosion, but also for composite repairs and related operational evaluations.

Aerospace NDT data fusion has more commonly been applied to failure investigations, and composite structure development and manufacturing programs, where highly flexible post-processing methods are more important than real time collocated processing methods. In a post-processing scenario, the need to provide unambiguous, provable conclusions is more critical than reducing data or image processing workloads. Parts may no longer be available for additional inspection, and the existing data may be from inspections that were widely spaced in time and physical location. In many cases, data presentations that are comprehensible by non-NDE experts such as manufacturing review board personnel or other investigative bodies are required.

Aerospace industry has also found a need for NDT data fusion in the development of advanced structures. The manufacture of modern complex aerospace systems has increasingly trended towards the development of large unitised composite structures, utilisation of precise process steps, and implementation of highly integrated non-destructive evaluation of the products, both during development, manufacturing, and in service. Where older generation metal based products such as missiles and aircraft used a uniform defect size specification, unitised composite structures have shifted to multiple mode NDE and zoned criteria fitness for service criteria, leading to unique requirements for NDT data fusion. Integration of NDT and composite fabrication process control measurements has proven to be an important element of product development and evaluation, and is often a critical technology factor in selecting composite structure design solutions.

At a fundamental level, the shift beginning in the late 1970s from designs based on traditional aerospace materials, with "standard" material properties, to composite materials, with "random", process-sensitive material properties has invoked a new requirement for multimode non-destructive material property measurement and characterisation. A number of Air Force and NASA major solid rocket motor failures and flight anomalies in the early 1980s (Westar VI, PALAPA-B, IUS QS-3) [1], spurred the aerospace community to address advanced composite NDE and material characterisation issues as a critical technology. This led to the implementation of a number of significant targeted research and development programs, and the consequent maturing of a number of related technologies, including NDT data fusion. These efforts were co-ordinated within the Joint Army Navy NASA Air Force (JANNAF) Technical Subcommittee processes, and were strictly controlled by US ITAR regulations because of the initial focus on rocket propulsion technologies. While some material specific sensitivity remain, aerospace companies are now applying the NDT data fusion methods and practices developed during this period to a wide variety of military and commercial aerospace applications. These methods have also been exposed to the academic and international community through public forums and media. The focus of these NDT data fusion methods is on maximising decision quality for unique, extremely high value components, rather than maximising evaluation speed.

In aerospace systems development, NDT data fusion is largely seen as a material and process and/or engineering tool, rather than a quality assurance technology, and this perception is reflected in the functional and operational deployment of the technology. In many cases, the objective of NDT data fusion is to convert combined NDT data into material properties for thermostructural and/or electromagnetic performance analysis. The processes require systematic engineering methodologies, which are often outside the norms of conventional NDT data interpretation. In particular, precise cross-registration techniques and techniques for normalising, cross-calibrating, and re-sampling are necessary. Of these, systematic registration of data with respect to the evaluated part is often a critical requirement.

The development of acceptance criteria based on individual part performance models requires precision multimode NDT data registration as well as accurate material property models and NDT data correlation. Since many very high performance aerospace materials are process sensitive, highly variable, and expensive, multimodal NDT analysis is also seen as a method to minimise the requirement for destructive testing in the development of material property data bases. By asserting similarity of materials in terms of multiple NDT signatures, the number of costly thermomechanical or other performance tests can be reduced.

7.2 BASIC THEORY OF NDE DATA FUSION FOR AEROSPACE NDT APPLICATIONS

To develop the relationships between NDE features and material properties, Boeing and other aerospace companies have supported the Air Force and NASA in developing a general purpose data manipulation methodology [2]. This one has been embedded in multiple releases of a government-owned software product referred to as the Integrated NDE Data Evaluation and Reduction System (INDERS) [3-6]. The philosophy behind the multiple incarnations of INDERS draws from finite element interpolation and integration theory, which simplifies comparing of multiple-modality NDT and mechanical test data for complex curved structures typical of modern aerospace applications. NDT data from arbitrary inspection points are logically extended to continuous analytic functions over space. This provides a systematic, statistically sound method for averaging, co-ordinate and raster transformations, dimensional reduction, and comparison functions by replacing heuristic approximations dependent on modality with uniform analytic algebraic operations, which preserve data integrity.

7.2.1 NDT Data Sampling

Advanced composite structures are characterised by continuous material property variations from point to point, which vary over a considerable range (compared with metals). For instance, interlaminar shear strengths of involute carbon-carbon structure may vary by as much as an order of magnitude within a single part.

Manufacturing improvements are constrained by complex processing steps and extremely long production flow time requirements that limit the improvements that can be achieved within time scales of planned missions. As a consequence, mission critical structures, such as exit cones, are non-destructively inspected using "best available" techniques and the data are used for determining the flightworthiness of each cone on an individual basis. In addition to visual and liquid penetrant testing, X-ray computed tomography; digital radiography, low frequency eddy current, and both pulse echo and through transmission ultrasonics typically are employed for involute carbon-carbon exit cone evaluation.

In order to relate NDT inspection measurements to material properties, such as strengths, stiffness and thermal expansion coefficients, destructive tests are performed on selected material samples. Relationships between NDT features and thermomechanical properties are developed. These correlations are employed in thermostructural calculations to establish margin-of-safety limits on NDT measurements as a function of location and distribution of material properties. Multiple modality inspection data are employed because of the considerable risk of system failure due to mis-categorising anomalous conditions using data from one modality. This is illustrated in figure 10-2, where part volume elements containing different material conditions but similar individual NDT signatures are compared in NDE feature space.

Mistaking a wrinkle for distributed porosity in an exit cone could result in the loss of a several hundred million dollar orbital deployment mission. Managing the multiple modality NDE data for complex shapes over different dimensionalities is a difficult practical problem. Typically, "whole part" inspection data are evaluated over volume elements corresponding to subsequently sectioned mechanical test specimens in order to establish statistically valid correlation between flight component NDT data and mechanical properties. It is often insufficient to perform NDT on the test specimen after excision, since the NDT data may no longer represent the critical anisotropies of the flight component such as ply and fibre orientations and edge conditions. Development of these multimode correlations with valid statistical norms (e.g. B-basis) is an example of a typical functional requirement for aerospace NDT data fusion processes. B-basis is a statistical requirement established by the US military and FAA for estimation of material properties used in the design of fail-safe structures.

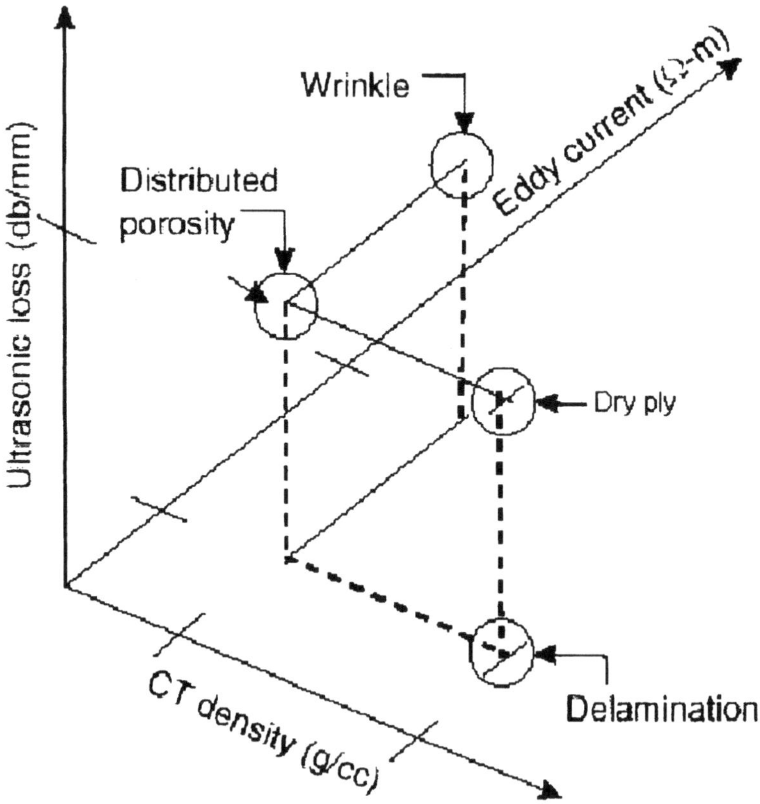

Fig. 7.2 Potential for single modality NDT to mis-categorise anomaly

The aerospace objective of estimating and reducing the variance in estimation of performance properties leads to a strict requirement for statistically sound NDT data sampling. Aerospace evaluations are very concerned with the provability of conclusions and the minimisation of risk, since failure of a critical component such as a nozzle exit cone would result in a minimum of US$150 millions loss, even neglecting the cost and schedule impact of loss of the payload. Sampling NDT data over the same volume element as a mechanical test gauge section is important for reliable property correlation to be developed. Studies in the 1980s showed that data registration and sampling errors were the largest source of variability in correlation between thermomechanical and NDE properties of high performance aerospace composite materials. Consequently, sampling and registration have been important foci of NDT data fusion methods.

Since the problem for composite materials is primarily expressed in terms of measurement of continuous properties, rather than identification of discrete defects, it is natural that methodologies for data handling focused on extending the inspection data to continuous functions, using the finite element modelling (FEM). A FEM based mathematical framework is used for expressing generalised NDT data manipulation operations, including interpolation, convolution, and dimensional reduction, freeing analysis of the data from dependence on the geometry or the scan raster of the inspection method. The mathematical framework used was borrowed from finite element methods for continuum analysis, which had not commonly been applied to NDT data prior to the mid 1980s.

For each inspection feature of interest, a mathematical generalisation of the feature extracted from the non-destructive inspection is defined. This generalised feature represents the instantaneous value of the NDI parameter at a point in space such that the weighted average over the domain of the inspection beam (or volume of interaction) would be the value measured and recorded in an inspection. Generalised features must have the property of being intensive, which is, the units in which the features are expressed do not depend on the geometry of the domain they represent. Examples of appropriate features illustrated in figure 7.2 are ultrasonic attenuation (dB/meter), CT density (gm/cc), or resistivity vector (ohms-meter). A measurable datum (such as the value of attenuation extracted from an ultrasonic waveform) is considered to be the weighted average over the volume of interaction of the generalised feature.

Following this step, an appropriate analytic functional form over space to represent the generalised feature distribution is defined. This can be done in several ways. If the physical principles of the interaction are understood well enough that the convolution distribution for the interaction can be prescribed analytically, then superposition of voxel influence functions can be employed. Alternatively, and more commonly, an approximating orthogonal set of functions is established, *a priori*, piecewise continuous over the domain of the part, and operated upon using a table of co-ordinates and element connectivity. The finite element theoretical approach to defining this functional form is most convenient, since it lends itself to being parameterised by inspection data, and eliminates dependence on a particular inspection raster. In this form, each node corresponds to the origin of a particular pixel (or voxel). The shape of the pixel (or voxel) is the distribution over space of the influence of the measuring interaction, rather than a discrete domain.

As an illustration, consider the distribution of a continuous physical property over a region. An NDT datum will be the result of the convolution of the "true" physical data over the shape of the sampling volume. This shape is usually expressed as a spatial function (e.g. point spread function), and can be measured by an NDT calibration approach. Linear interpolation between NDT sampling points (using an automatically generated finite element mesh) allows us to approximate the intensive physical property at any point in the region. Common volume element mixing or comparisons can now be made, by analytically integrating the continuous, piecewise linear function representing the data over the chosen set of common volume elements. If a series of infinitely small volume elements (points) are selected as the target volume elements, the result is simply a linear interpolation of the data. If the target volume elements correspond to a performance model's finite element mesh, the appropriate average of the data over the finite element is the result.

Complementary mathematical approaches to evaluate inspection features over arbitrary volume, surface, line, and point elements are also developed. Generally, evaluating an inspection feature for arbitrary elements, such as voxels from other inspection modalities or mechanical test specimens, becomes a matter of numerical evaluation of domain integrals, which by employing conventional finite element theory can be reduced to algebraic expressions in terms of the original inspection data.

Typically, a software framework for handling the data computationally in a unified and flexible form independent of the scan raster is established. Once available, the operational steps for data reduction and analysis for multiple modality inspections for a known part is reduced. Typical functions of the software frameworks are surface and volume domain averaging, co-ordinate and raster transformations, dimensional reduction, and comparison functions. By avoiding heuristic approximations dependent on modality with analytic algebraic operations, NDT data is handled in a statistically sound manner and is readily transportable into FEM analysis.

7.2.2 NDT Data Registration

An essential element of NDT data fusion is the registration of the data to part fixed co-ordinates, which involves mapping the data to a geometry model of the part. Using finite element definition of the geometry allows the NDE information to be readily available for other uses such as structural analysis.

Since the unique characteristics of composite materials propelled industry into the development of continuous property theory and methods based on FEM, it was natural that other theoretical benefits became available. It was recognised that functions of spatial location (based on Eulerian co-ordinate systems) associated with most NDT inspection systems were not suitable for accurate part registration, since part geometries would often change significantly between NDT processes (e.g. due to additional process steps, different gravity or fixture loading, etc.). To solve the problem, the development of methods for registering, transforming, and manipulating data in Lagrangian (part fixed) co-ordinate systems was needed. This technical issue had long been solved in FE practice, by requiring explicit co-ordinate and connectivity tables, and provision of algorithms for the transformation of data models between Lagrangian and Eulerian frames of reference.

Lagrangian co-ordinate systems are defined based on knowledge of the part geometry, design, function, and process steps. Conceptually, Lagrangian measurements are those locating data that can be used by a technician using chalk and string on a physical part. For instance, on a rocket motor exit cone, multiple Lagrangian co-ordinate systems can be defined which express position in terms of several meridionals, circumferential, and through-thickness positions relative to a physical fiducial. Mathematical transforms between the differing Lagrangian frames (e.g. "inside surface meridional position" to "outside surface meridional position") must be established. Since 3-D CAD models of aerospace part geometries are routinely available, the development of these transforms is relatively straightforward for an engineering analyst even if the part is not available yet. General practice at Boeing is to embed the transforms in software or in symbolic expressions that can be parsed and executed dynamically from within the NDT data fusion-operating environment.

To provide data fusion functionality for the unique priorities of NDT data decision processes, NDE data processing algorithms that deal with the geometry associated with parts and inspection systems are needed. In mathematical terms, this means being capable of transforming data from either Eulerian or Langrangian co-ordinate systems into a Lagrangian (part fixed) system before combining data. Instead of navigating in an Earth or camera based geometry system, NDE data fusion algorithms must be capable of navigation within a part based geometry system to be effective. Implementation of part fixed co-ordinate registration is the key functional requirement for NDE data fusion and is the fundamental element of the INDERS.

Part fixed co-ordinate registration is most easily implemented by utilising finite element modelling concepts for attaching geometry information to NDE test data. In the INDERS implementation, the target of NDE data conversion processes is the unstructured cell data (UCD) format [7], an extremely simple ASCII data format standard that supports the finite element model for geometry information. Once NDE data, measurement data, and part geometry data is converted into UCD format, software tools are provided which permit transformation of data into part fixed co-ordinates and, consequently, a wide variety of NDE data fusion processing steps. The data can also be visualised and transformed dynamically using a variety of computer visualisation interfaces.

7.2.3 NDT Data Mixing

If data are to be mixed, the mixing logic is applied only to the correctly registered and resampled data that results from the evaluation of raw NDT data onto a common volume element discretisation. Often, this volume element discretisation is the finite element mesh of the performance model for the part being tested, and the mixed NDT data feature is and estimate of strength, stiffness, thermal, electromagnetic or other performance related property. This volume sampling requirement on data mixing grew out of a number of studies which found that the wide scatter in single mode NDT to mechanical property correlations for composite materials was greatly reduced by carefully sampling the NDT data over the gauge section of the mechanical test.

Comparing data from multiple modalities involves the application of these same techniques. An example of data from three NDE techniques taken from the same exit cone is presented here. Figure 7.3 shows features from computer tomography (CT), ultrasonics, and eddy current after they have been mapped onto an equivalent part-fixed co-ordinate system. The three-dimensional CT data has been averaged through the thickness of the cone wall and evaluated on the ultrasonic pixel raster (in this case, the voxel set with the largest elements). Similarly, the eddy current data has been interpolated to the ultrasonic pixel raster. The three techniques can now be compared for evidence of correlation to material property changes or each other or for mixing on a statistically sound basis.

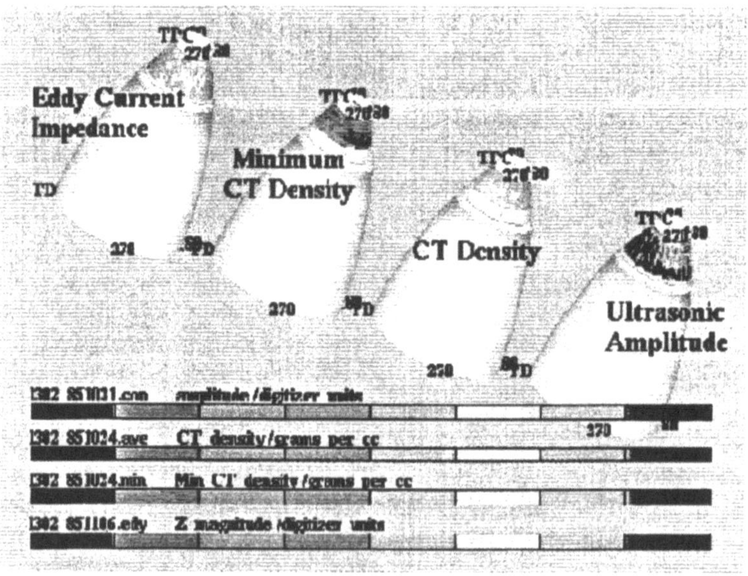

Fig. 7.3 Visualisation of co-sampled, co-registered exit cone NDT data

This framework for managing multiple modality NDE and associated process and destructive test data is sufficiently general to be applied to a broad range of part configurations and NDE features. The key advantage of this approach is that all of the data is managed in a consistent paradigm, freeing the analyst to spend his time analysing the data, rather than managing it. Because the approach is general, it has applicability to non-NDE problems, such as problems requiring fusion of data from in process monitoring and destructive testing.

7.2.4 The INDERS System

Visualisation is the art and science of turning complex data into visual insight. Since one third of the human brain is devoted to visual processing, providing data in visual form has the potential to increase comprehension rate significantly. In computer visualisation, the power of the human eye and brain are used together with the computer's data processing power to permit rapid comprehension of complex data relationships by presenting data as multi-dimensional colour images and animations.

Figure 7.3 shows the integration of an inertial upper stage (IUS) exit cone geometry with eddy current, ultrasonic, and X-ray CT inspections of an IUS SRM-1 rocket motor nozzle exit cone as a three dimensional representation. The view can be manipulated dynamically using a computer mouse. More importantly, it preserves the actual geometric relationship between the test data and the exit cone shape. A simple question like "how far is that anomaly from the compliance ring ?" can be determined as if the actual part were sitting in front of the analyst with the data chalked on. In fact, the system used in this illustration provides for the use of digital chalk, lines, and rulers.

Essentially the data acquired from process monitoring or NDE operations are expressed in a part fixed co-ordinate system using techniques commonly used in finite element analysis. Features of interest can be calculated by appropriate combinations of inputs from the various sources that are transformable in a common co-ordinate system. The results can be mapped back onto any of the original test system representations or presented in standard computer graphics, visualisation or data base representations. The INDERS software package, has won acceptance on various military and commercial programs, and has been used by these programs to support studies, investigations, manufacturing process development, and qualification activities.

The integrated approach implemented in INDERS to handle non-destructive evaluation and process control measurements is a useful approach for the review, evaluation and disposition of processed articles, particularly large unitised structures. The fundamental requirement for NDE data fusion is the co-registration of data obtained from various techniques or at different times and locations in part fixed co-ordinates. This allows mapping the data to a geometry model of the part and allowing that data to be reviewed in a single workspace, increasing confidence in accept or reject decisions. Finite element definition of the geometry data is the fundamental building block enabling NDE data fusion and allows the NDE information to be readily available for other uses such as structural analysis. The implementation in INDERS of a visual programming methodology with an object oriented tool set allows rapid creation of new data handling and comparison schemes. The ability to visualise both process information and reduced inspection data on a part fixed co-ordinate system is critical to the interpretation of the quality of modern structural products and greatly aids the development of product acceptance criteria.

The visual programming interface, referred to as the network editor (NE), is used for programming applications. The network editor allows users to construct visualisation applications for NDE data fusion as connected, hierarchical networks of objects, with drag-and-drop convenience. This approach promotes software reusability and increases programming productivity. With visual programming, a non-expert user can create, modify, and combine program objects into higher-level application objects. By displaying the hierarchy of programming objects and their relationships visually, the user is encouraged to use a structured approach to application construction, which dramatically shortens the time it takes to develop, test, and deliver NDT data fusion applications.

The basic implementation strategies behind state-of-the-art computer visualisation tools are the finite element modelling concepts. In visualisation practice, these concepts have been generalised and extended from the relatively simple integration and interpolation methods used in engineering analysis to include such functionalities as texture mapping, light modelling, depth cueing, and high order interpolation elements such as non-uniform rational B-splines (NURBS).

Boeing uses INDERS as its analysis and visualisation tool and UCD as its visualisation model file format. However, models produced can be converted to a number of different visualisation model formats (e.g. VRML) and associated visualisation platforms. INDERS visualisation is based on AVS/Express, a multi-platform, component-based software environment for building applications with interactive visualisation and graphics features. It employs an object-oriented visual programming interface to enable the user to create, modify, and connect application components. In addition, it provides a wealth of fine-grain visual programming objects that provide a complete development environment. High-level objects are available, such as 2-D and 3-D graphics viewers, data and image processing algorithms, and graphic user interface tools.

INDERS will operate on a variety of platforms including Silicon Graphics workstations running IRIX and PC workstations running Windows 95 or Windows NT with at least 32MB of memory.

The most recently released version of INDERS (version 3) was demonstrated to government and industry at the March 1998 JANNAF Technical subcommittee meeting in Salt Lake City. Presentations, hands on tutorial, and the runtime package were distributed on CD-ROM.

Government and industry members of the JANNAF NDE subcommittee met at the 1992 JANNAF NDE Data Fusion Workshop. The meeting resulted in a draft recommendation for NDE data fusion technology development by the Air

Force. This recommendation formed the basis for the NDE data fusion contract effort. The Air Force's contractors used a series of usage scenarios to assist in the final definition and prioritisation of NDT data fusion workstation requirement elements.

Among the key conclusions of the requirements analysis was the recognition that *NDE Data Fusion* is an evolutionary process in which new requirements arise in response to new or unusual problems. This recognition led to a toolbox design concept rather than an end product design concept for the software component of the workstation. In this design concept, the NDT data fusion workstation is used as a factory for manufacturing NDE applications, suitable for both rapid prototyping of new applications and running of previously developed applications. This approach relies on modern object oriented programming (OOP) concepts which have replaced traditional (i.e. development of FORTRAN subroutine libraries) approach to providing reusable software tools.

As a consequence, INDERS is a layered collection of software tools tailored for NDE data analysis and operating within the interactive programming and visualisation environment AVS Express. These tools were developed for high-end Silicon Graphics workstations, but are platform independent. They include data fusion tools that allow rapid development of NDE data fusion applications.

It is convenient to think of INDERS 3 as a factory for manufacturing NDT data fusion applications. The first layer is very rich in general purpose software tools, is used to develop NDE specific tools, and requires a software engineer with a significant amount of training to use. The second layer is tailored for the NDE engineer, has much fewer, highly integrated software tools just for NDE data analysis, and requires a modest amount of training. It is in this layer that the NDE applications are built. The third layer is the end user application. It is designed to be very easy to use, has only the tools available necessary for the application, and requires very little training to use. Anyone familiar with computer graphic user interfaces such as pull down menus and mouse operations can use applications. Developing applications requires some knowledge of visual programming using AVS Network editor. Developing new software components requires a software engineer who has completed the AVS/Express developer training provided by the Advanced Visual Systems Inc.

7.3 AIMS AND OBJECTIVES OF NDE DATA FUSION IN AEROSPACE APPLICATIONS

Since the advent of digital NDT data acquisition systems, NDE researchers and practitioners have envisioned reducing large quantities of NDT and part geometry data into a unified model for convenient evaluation by non-technical end users. Both post-process NDT data fusion and sensor-based NDT data fusion have been identified as an important aerospace technologies with demonstrated cost benefit for evaluation of high value assets, such as rocket motor and military aircraft components. The post-process technology has potential for reducing ambiguity and improving decision-making processes in a wider context.

The primary objectives of integrating NDT data with part geometry are to reduce time for non-experts to understand complex NDT data. In cases where complementary sets of NDT data have been acquired for a single part in different frames of reference (e.g. on different NDT systems), an additional objective is to permit cross-registration of the complementary data, a prerequisite for post-process NDT data fusion. In addition, the drive to more affordable manufacturing is leading to complex, monolithic structures and graded fitness-for-service criteria that require NDE data and geometry integration capability. For most applications, data and geometry integration is associated with three-dimensional visualisation functionality, which has become widely available on personal computers.

Integration of NDE data with geometry in a visualisation environment is applicable and/or cost beneficial in a number of areas. In addition to supporting high value component evaluations, it has a major role in developmental programs, in failure investigations, in NDE science, and in the evaluation of complex monolithic structures. The drive to affordable manufacturing is leading to an increase in the development and production of complex, monolithic structures requiring NDE data and geometry fusion functionalities. NDE data fusion technology is increasingly becoming a key requirement in these development programs, and is expected to be used during production. NDE data fusion will be applied to post-deployment and maintenance scenarios such as radome refurbishment where mission criticality is an issue. Low observable signature assurance is another area that falls into this category. Because of prior work, the cost of introducing this technology to program practice has been substantially reduced.

NDT data fusion was first implemented in aerospace applications to support manufacturing review boards and failure investigations associated with the development and deployment of advance composite structures. The technology has often focused on establishing quantified margins of safety on unique, very high value components through the integration of NDT data into structural analysis models. A more recent goal has been to transfer NDE data fusion technology developed on prior Government and industry programs to Air Force depot operations and maintenance facilities. Aerospace industries, such as Boeing, are also deeply involved in supporting the operations and maintenance (O&M) needs of the Air Logistics Centers (ALCs), both as a direct support subcontractor and as a technology provider through various Air Force Laboratory programs. Although current applications are immature, a goal of aerospace NDT data fusion is to achieve significant savings in the O&M arena by:

- eliminating unnecessary disassembly, repair, and replacement costs due to diagnostic ambiguity,
- enabling the unambiguous interpretation of emerging low cost imaging inspections methods,
- and reducing the skill level and training requirements for reviewing inspection data.

Boeing has identified specific applications in aircraft maintenance where savings are realisable, and has performed a number of studies to illustrate the applications.

At a time when the ALCs are downsizing, the repairs and inspections required to maintain the Air Force's ageing fleet is increasing. Many military aircraft are operating well beyond their original design lives, and improved NDI/E is one of the key requirements to keep them flying. A potential high payoff improvement is to employ improved methods for analysing and managing NDT data, such as NDT data fusion methods. Combining different modes of NDE data from the same location or combining more than one set of data from the same mode, such as the data on a part taken at different times can result in a significant cost benefit.

The payoff to NDT data fusion is usually a consequence of the fact that critical features of the part being inspected can be much more positively identified when the data are fused than when used independently. In some cases, data fusion is the only way to identify a critical feature of the part. Data fusion has been demonstrated to reduce both positive and negative false calls, improve image interpretation, improve reliability of inspections, and lead to a more quantitative assessment of the part being inspected. In selected applications in the rocket motor industry, it has lead to millions of dollars in cost savings. The data fusion process usually requires more effort than dealing with the un-fused data, so trade studies are needed to determine if the payoff justifies the cost for each application. It is important to include the effect of data fusion on the entire maintenance effort, and not just the time to conduct the inspection. In general, this requires exposure of the end-to-end process benefits to higher levels of management.

Even with the advantages of NDI/E data fusion demonstrated, implementation of NDT data fusion in the operations and maintenance environment has progressed slowly because of several factors. The first factor is that many people in the aircraft maintenance community are unfamiliar with the full potential of NDT data fusion developed in the rocket community. A second factor is the lack of clear identification of the applications for which there is a net benefit within maintenance and operations, and demonstrations of the effectiveness of the approach. Identification of such applications has been a key objective of a number of R&D efforts. A third factor is that the payoff must be presented in a form relevant to the end user for him to be convinced that his investment in this new approach is justified. The importance of this third factor is frequently underestimated, resulting in high quality technology too often not being used. This is why NDT technologists have placed a heavy emphasis on the case studies of potential high payoff applications. A fourth factor that can impede acceptance of new software is that the interface between the end user and the technology is too complex. One of the key problems is transitioning highly analytical process from MRB decision making to NDT practice where a level II inspector can use the technology comfortably. A final factor is organisational or cultural. For example, an NDT manager will be reluctant to embrace a new technology if he is penalised for an increase in the flow time through his organisation, even if it results in a net improvement in the total process.

7.3.1 Mechanical Property Correlation

The widest and most common aerospace application of NDT data fusion is in the area of predicting mechanical properties of advanced composite materials. Medical CT equipment and industrial ultrasonic systems were widely used to evaluate carbon composite materials.

Several programs were conducted to correlate NDE measurements with mechanical data. The programs consisted of developing data collection and reduction techniques, performing detailed component measurements using both the ultrasonic and CT systems, performing mechanical properties tests and performing data correlation analysis. These programs helped establish the methodology for data exchange and data correlation that was used on operational programs such as IUS. Establishing statistically sound evidence of correlation between accurately calibrated CT density, pulse echo based ultrasonic attenuation measurements, and mechanical property measurements was the chief concern of these efforts.

On IUS, concern for the integrity of carbon-carbon (C-C) cones grew following flight and test-firing failures in various solid rocket systems. Further, NDE data collected on several IUS exit cones were anomalous; however, cause for rejection could not be easily established. The integrity or thermostructural performance and parameters measured by NDT had to be related.

A scan criteria program [8] was conducted to establish a database of NDE parameters and of strengths and other properties, as well as to study the relationship between NDE parameters and structural response.

The objective of this study was to quantify the mechanical properties of exit cone material that has variable ultrasonic and CT quality. The mechanical property measurements of local areas, identified as defects, were then represented in exit cone structural analyses to determine the effect on exit cone structural margins of safety during motor firing.

The scope of the study included performing, documenting and correlating the results of ultrasonic and CT scans inspections of available fired and unfired exit cones. Uniform and non-uniform areas of exit cones were selected to furnish material for mechanical testing of exit cone material. Exit cone material was mechanically tested to characterise properties relative to ultrasonic or CT scan features. Mechanical tests were run at temperatures over the operating range of the exit cone during motor operation. The mechanical properties of the material were estimated from NDT data and inserted into structural analyses of the exit cones for temperature and pressure conditions during motor operation. The fired exit cones were analysed using the observed CT scan and ultrasonic conditions and the correlated mechanical properties at the observed locations in the cones. Margins of safety were reported for all anomalous regions evaluated and for all failure modes.

The basic approach used during this program consisted of generating an NDT database from materials that exhibited a wide range of NDT features, subjecting the material to a wide range of mechanical property tests to establish a mechanical property database, establishing the degree of correlation between the two data sets, performing structural prediction calculations using various material parameters, and finally recommending a multimode NDT accept/reject criteria.

The IUS scan criteria program has been the model for subsequent Boeing applications of NDT data fusion for unique composite parts. Currently, Boeing is applying a similar approach to evaluation of composite patch repairs for ageing B-52 aircraft. Since each patch has unique geometry and loading, individual multimode NDT-based structural analyses must be performed to qualify them.

7.4 BENEFITS AND LIMITATIONS OF APPLYING DATA FUSION TO AEROSPACE APPLICATIONS

In July 1992, an NDE Data Fusion Workshop was held at the JANNAF NDE subcommittee meeting. There was a consensus of the aerospace NDE community attending this workshop that NDE data fusion capability was becoming an increasingly important requirement, but needed further development to be accessible to the aerospace NDT community. As a result of recommendations from this workshop, in November of 1994 Air Force Research Laboratory (AFRL) issued a program research and development announcement to develop an NDE Data Fusion Workstation. Boeing won the contract to develop the workstation, and demonstrated the workstation at a JANNAF workshop. In this workshop, the focus was to assist participating aerospace NDE technologists understand how to take advantage of the investment made by AFRL to help do their jobs faster, better and cheaper.

NDT data fusion faces the common limitations to new computer-based technology insertion, which are lack of knowledge, lack of payback information, lack of relevance, and poor user interface. The calculation of cost benefits is difficult. Benefits may involve direct cost savings or offer indirect savings, such as risk reductions that are difficult to quantify financially. The direct cost savings can include reducing the overall capital cost of NDT equipment by replacing data processing functions performed on separate equipment, reducing operating cost by centrally locating the data and processing functions, reducing labour and accelerating schedule by increasing processing and data interpretation speed, and reduced training by having a common workstation for processing NDT data from a variety of modes in a consistent and uniform representation. A frequently overlooked direct cost savings is the reduction in meeting time and management effort when data interpretation ambiguities and/or controversies arise. An NDT data fusion workstation, by providing less ambiguous information about a component, will significantly reduce or eliminate the effort expended in such meetings. The reduced schedule time benefit not only saves at the inspection organisation level, but also increases overall system readiness - a critical cost factor for the Air Force and other aerospace customers. The reduced risk benefits may include more accurate measurements, increased confidence, improved knowledge, and higher reliability. These benefits are commonly realised through a reduction in false call rate for inspections. In the case of aircraft maintenance, the Air Force expends over US$10 billions per year of which NDT represents about 5% of the cost. False calls in many inspection areas can run as high as 10%. Reducing false calls in half (from 10% to 5%) would have a cost impact of US$25 millions per year. If NDT data fusion technology is applicable to only 10% of the NDT operation, it can be expected to offer potential savings of over US$2.5 millions per year. Of greater significance is the fact that a critical false call could result in substantial costs through unnecessary repair effort or possible failure during missions. As a result, small improvements in reducing false calls can result in large savings for the Air Force.

7.4.1 Cost/Benefit Issues

A prerequisite for insertion of improved or advanced technology such as NDE data fusion into aerospace component maintenance inspection and material NDE is a perceived benefit. The benefit of the new technology may be simply a monetary cost saving as an alternative to more expensive approaches currently employed. Often, however, it is a more complex saving involving schedule acceleration in the short or long term and reduction of risk through improved assessment of the component condition. These benefits can be difficult to quantify. An effective method for insertion of new technology into the work environment, when benefits are complex, is to provide demonstration case studies which show the benefits of the new approach in a manner easily appreciated by interested parties. The aerospace industry has found this approach particularly effective in the other technology insertion programs. From case studies, management and technical personnel are able to extrapolate the benefits of the technology to their specific applications. The case studies demonstrate the real world application of the technology, how it is applied, how the results are used, and the ensuing benefits to the specific program and overall organisation. By selecting case studies that involve components or problem areas with real world significance, the implementation of NDE data fusion can be properly directed.

Boeing's experience with Air Force and NASA missiles and space programs has shown that, in selected application areas, utilising NDT data fusion produces very significant benefits. In addition to the Air Force Inertial Upper Stage program's cost savings of US$11 millions from exit cone recovery and more than US$150 millions for avoidance of an in-flight failure, the NASA Productivity Enhancement Laboratory at Marshall Space Flight Center reported that the implementation of Boeing's data fusion software at the laboratory in 1989 reduced the flow time for evaluation of carbon-phenolic nozzle components from 30 days to less than 2 days.

NDT data fusion remains applicable and/or cost beneficial only in specialised areas. In addition to supporting high value component evaluations, it has a major role in developmental programs, in failure investigations, in NDE science, and in the evaluation of complex monolithic structures. The technology pays off in drastically reducing the time to develop NDT data applications, improving ability to comprehend NDT data, and reducing skill required developing NDT applications. It also greatly improves ability to quickly prepare compact readily comprehensible presentation material, leading to improved communications

7.4.2 Geometry Modelling Issues

The visualisation of information, such as NDE measurements, about a component are best understood within the context of part geometry. Therefore, an important aspect of data interpretation is transforming the data into part fixed co-ordinates and assembly of data into a visualisation model which is integrated with the geometry of the part. Integrating geometry models with measurements is one form of data fusion. In recent years, a number of tools have been developed by aerospace companies to assist in the handling of multimode NDT data. The primary activity is the conversion of the data into a fully integrated visualisation model, using a data object description format such as UCD. The UCD format describes both data and geometry using well established finite element approaches, allowing data to be manipulated and displayed in 3 dimensional visualisation packages.

Part geometry information is acquired from CAD models of components or, in some cases, reverses engineering methods, such as CT and laser ultrasound. For complex geometry parts and for data obtained from multiple NDT modes, this capability provides a necessary correlation for correct interpretation of results. In addition, this approach supports exporting of NDT data into other analysis environments, permitting analyses of as built parts containing flaws of other anomalies.

Three-dimensional CAD methods and associated computer visualisation systems have become extremely important elements of aerospace define and manufacturing processes and are essential for competitors in the aerospace marketplace. Three dimensional part geometries are a major component of most CAD models, and are often rendered from very high order CAD descriptions into simpler modelling descriptions for a variety of purposes such as thermal, structural, and NDE analysis.

By expressing part geometries in UCD or another equivalent format, they become logical entities much like NDE data. While they are not in the frame of reference of a particular NDE inspection system, co-ordinate transformations can be applied to either the data or the part to put them into a common reference frame for visualisation. Usually, NDE data is transformed to the co-ordinate system of the part geometry, or both are transformed to a viewer co-ordinate system.

Geometry models have many other uses in NDE practice. For instance, replacing forged or built up structural members by castings may result in large cost benefits. However, low cost cast components often vary significantly from original design configurations and require methods to capture the as built geometry, to prove inspectability, and to establish zoned fitness for service criteria. While these tasks can be achieved by building castings with controlled defects and performing X-ray or other NDT, aerospace companies are increasingly utilising lower cost NDT simulation methods together with accurate geometry models of the casting for satisfying these requirements. Often, the accurate geometry models are the products of NDT data fusion processing.

NDT techniques such as CT and laser ultrasonics (LUT) are frequently being used for capturing as built geometries for manufacturing tools (e.g. drills), castings, obsolete parts, and ergonomically developed parts (e.g. helmets, joysticks). Heritage programs need to capture casting geometry into CAD models to enable designers to package additional components into existing systems, reducing the cost of system upgrades. CT reverse engineering methods are increasingly recognised as general-purpose versatile methods for 3-D geometry model acquisition. Existing CAD models provide a starting point for derivative geometry models such as for replacement castings. Figure 7.4 illustrates models for NDE analysis, which were built by fusing a sectioned CATIA model with X-ray CT and physical measurement data for a drop link frame.

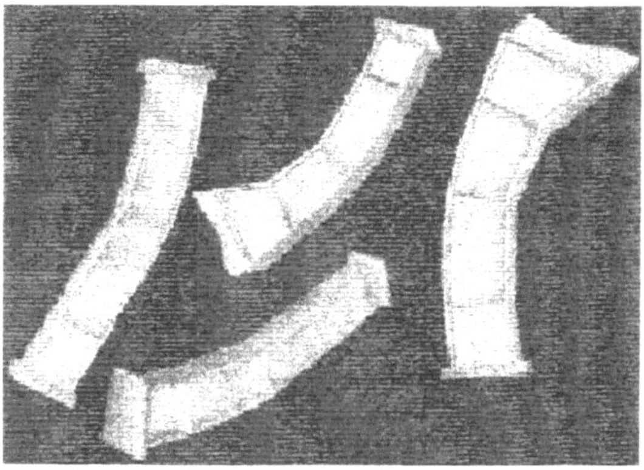

Fig. 7.4 Casting geometry models fused from CAD, X-ray and measured data

7.5 NDE DATA FUSION FOR AEROSPACE APPLICATIONS

7.5.1 Rocket Nozzle Exit Cone Evaluation

Carbon-carbon rocket motor nozzles are very high value items because of their cost to manufacture, their long schedule time, and the extremely high value of their payload. Multiple mode NDT techniques is applied to rocket motor nozzle exit cones to assess both, as manufactured quality and flight worthiness after storage and handling. Multimodal methods used include ultrasonics (UT), eddy current (EC) and X-ray computed tomography (CT). Each inspection modality (e.g. CT or UT) can indicate the presence of an anomaly without necessarily distinguishing the material defect. For example, CT images of a dry ply and a delamination on two different parts are essentially indistinguishable. Conversely, different NDE modalities may respond to an anomaly to differing degrees. This concept of using several modalities to distinguish an anomaly is generalised into a multi-modal parameter representation of anomaly types illustrated in figure 7.2.

To correctly identify manufacturing defects and damage, it is critical that the data from each mode be viewed in precise co-registration. Figure 7.3 shows how data from ultrasonic, computed tomography, and eddy current can be presented together in part fixed co-ordinates. The rocket nozzle exit cone shape is shown as a surface with features from each methodology displayed over the critical (i.e. highly stressed) forward region of the nozzle exit cone as shown. The quantitative parameters evaluated here are eddy current impedance, minimum CT density through the wall, CT density averaged through the wall, and ultrasonic amplitude. Low density CT indications can be due to wrinkling, dry ply conditions or delaminations. The "dry ply" condition can have sufficient across-ply tensile and interlaminar shear strengths to support the mission loading. A delamination or wrinkle, however, would result in mission failure if misdiagnosed as a dry ply.

Only by considering complementary features of ultrasonic and CT data can the two conditions be distinguished. In this example, the combination of anomalous data from ultrasonic and eddy current scans allow a correct interpretation of an otherwise benign low density indication in the CT scan as a wrinkle whose presence would almost certainly lead to flight failure.

7.5.2 Commercial Aircraft Drilling Process Optimisation

Figure 7.5 shows the fusion of a CT reverse engineered geometry model with CT derived thickness for a chip from a high speed drilling operation. In figure 7.5, P is the co-ordinate representing the Lagrangian position, measured along the path represented by the outside edge of the chip in mils (thousandths of an inch). N is the co-ordinate representing Lagrangian position along the chip in a direction normal to the path traced by the chip edge, also in mils. The chip is a witness item that encodes in its thickness the critical drilling parameters being studied. These are used to calibrate high temperature, high strain rate 3-D drilling analyses which are used to develop low cost, high speed manufacturing processes. By unwrapping the chip into a time registered cutting history, very high frequency drilling dynamic effects are resolvable, eliminating the need for expensive dynamometry testing.

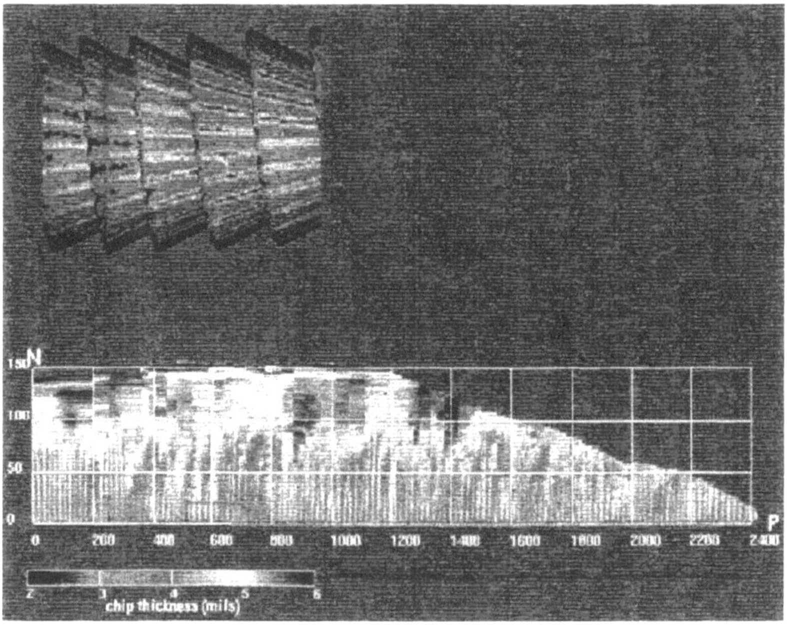

Fig. 7.5 Lagrangian transformation and fusion of CT chip geometry data

7.5.3 Radome Coating Performance

The performance of certain radomes containing sensitive instruments such as radar systems is critically dependent on its paint and paint thickness distribution. Inspection of the radome for paint thickness is an essential NDT activity. The measurement of the thickness is obtained ultrasonically. An example of differential ultrasonic thickness mapping is displayed on the geometry model of the radome as shown in figures 7.6 and 7.7. In this example, the thickness distribution after an unsuccessful and successful radar test qualification is subtracted to identify areas where the controlled painting process was deficient.

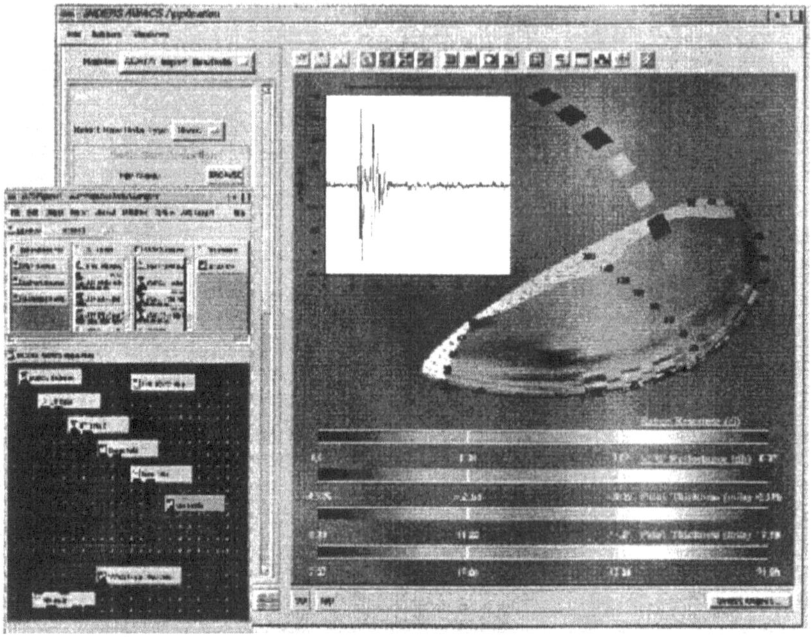

Fig. 7.6 Visualisation of multimode NDT and performance data for radome

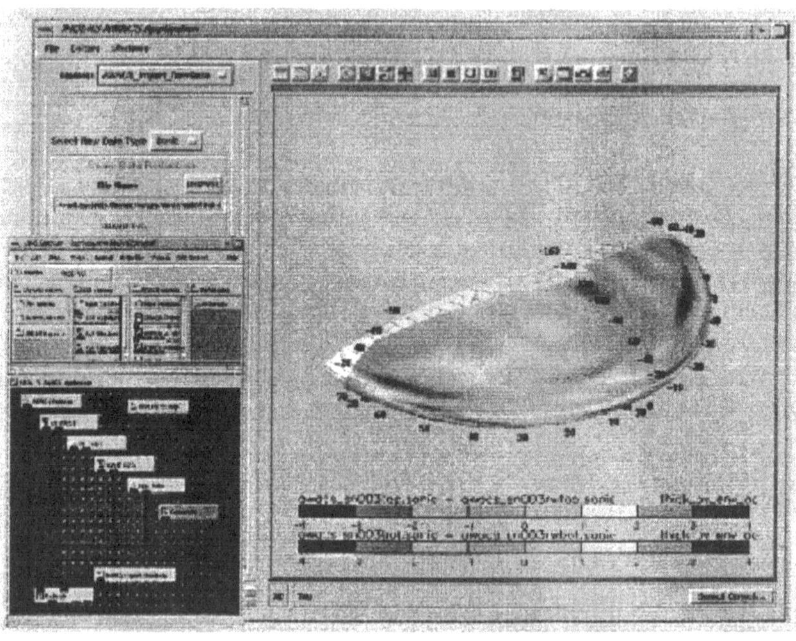

Fig. 7.7 Differential ultrasonic thickness registered to radome geometry

All components of the corresponding radar range data and other NDT data can be displayed in a similar geometry-registered format by selecting the appropriate GUI options. Using these measurements and the display capability in a single workstation allows improved interpretation and disposition of radome manufacturing and repair issues.

The study that led to the operational radome repainting process demonstrated the fusion of radome geometry data, full waveform ultrasonic inspection data, microwave (horn stethoscope) reflectance data sondicator rake, and full field radar range test data. Figure 7.6 shows a visualisation of the radome together with raw ultrasonic waveform data, the ultrasonic thickness derived from the raw data as a colour mapping, superimposed microwave reflectance measurements as a panellised colour mapping (near the equator of the radome) and selected radar response data as a function of the azimuth angle for a selected frequency and receiver orientation. Eliminating a single radar range test for the radomes saves approximately 600 man-hours required for radome range testing. The application provides a method for examining the relationship between paint thickness, substrate configuration, and radar performance. Figure 7.7 shows a difference between the ultrasonic thickness after first strip and repaint which resulted in a radar test failure and the second strip and repaint, which resulted in a radar test pass. These visualisations provided the program with immediate insight into how to reprogram the painting robot to eliminate the expensive re-work associated with stripping, repainting, and re-testing on the radar range.

7.5.4 Induction Welded Thermoplastic Composite Structure

Another example of data fusion for a complex process is the integration of process monitoring data and NDT measurements for thermoplastic induction welding [9]. This process uses a metallic susceptor between thermoplastic surfaces. The susceptor is inductively heated causing the surfaces to fuse. Time, temperature, and pressure are the critical parameters that must be sufficient for welding but not over processed to cause voiding in the skin or cap composite structure. During welding the time and temperature are monitored at specific locations in the weld zone using thermocouples and may be adjusted by the induction coil controls. The time at temperature for the weld interfaces has been determined to be the most important measurement for the process control. The pressure to the tooling is a fixed value assigned for specific welding conditions.

The temperature data is obtained from thermocouples placed at various locations along the weld and also at either the spar surface or between ply layers around the susceptor. The equivalent time (in minutes) to a process target temperature is calculated for each thermocouple and can be displayed as a function of position along the weld.

Ultrasonic inspection of the weldline is performed at various stages of the welding process. During development, UT scans have been taken at various times in the process to understand the state of the welding and its progress. The most reliable inspections, however, are after the welding is completed and are taken with a full waveform digitising scanning system. This full waveform data is processed to compare signal strengths from the weldline and the back of the spar cap signal for an ultrasonic pulse echo inspection taken from the skin surface.

Figure 7.8 is an illustration of combining test data from both the final ultrasonic inspection and the processing time at temperature. The visualisation shows the geometric configuration of a welded skin to a T spar. The geometry data includes chalkmarks for position information. The time at temperature data are shown at the top of figure 7.8, displayed by coloured panels at the position of the thermocouples in the test object. The depth locations in the object are difficult to distinguish in the figure, but can be shown by manipulation of the visualisation display. Ultrasonic models shown in the lower portion of figure 7.8 are processed image data overlaid on to the geometry model of the weld. The bottom model is the ratio of the weldline echo to the front face surface echo. The middle object is a weld factor calculation which is a point by point calculation of the ratio of the difference of the back of spar cap signal to the weldline signal to the sum of the back of spar cap signal and the weldline signal. This value is positive for a good weld and negative for a poor weld. This weld factor model is processed data that essentially fuses ultrasonic response data information from different depths in the part.

To a skilled process engineer, figure 7.8 shows a data fusion visualisation of an induction welded test sample. The display includes visual models of process thermocouple data, weld quality factor extracted from the ultrasonic signals, and the weld interface response, all overlaid on geometry models of the test sample. The user of the application can call up a full waveform ultrasonic viewer and can revise the signal feature extraction parameters for recalculating full waveform ultrasonic signal feature maps. In addition, in process MAUS data and other signal features of Boeing's large blade/fillet UT scanner can be called up and displayed on any of the geometry models.

The NDE data fusion workstation has become an integral part of Boeing's affordable composite process development strategy under the Composites Affordability Initiative (CAI) program. The CAI program is using an INDERS-based NDT data fusion workstation for fabrication and assembly process development. Several related Boeing manufacturing development efforts, including the wingbox development and a titanium casting initiative also are using a NDT data fusion workstation.

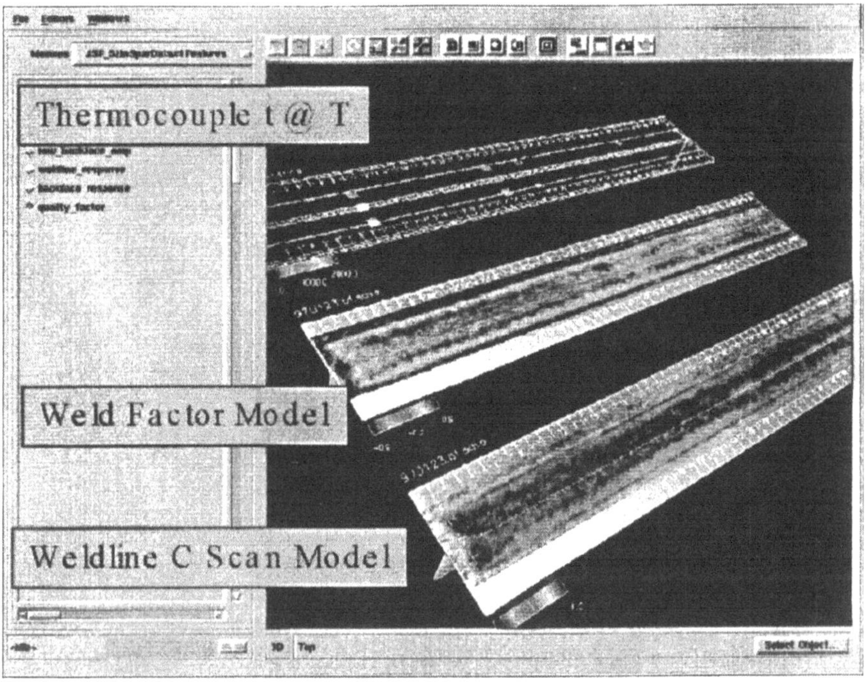

Fig. 7.8 Induction welded T-spar data fusion application interface

7.5.5 Other Applications

Boeing has used the NDE data fusion workstation on a number of other programs. A large number of applications were developed in support of low cost manufacturing initiatives. In general, these efforts were focused on fusing NDT data with geometry models and/or measurement data. In addition, the NDT data fusion workstation was used to support radiography simulation, ultrasonic simulation (both conventional and guided wave), NDE data reduction, and a variety of image processing and analysis tasks. Figures 7.9 and 7.10 illustrate some of these applications.

Figure 7.9 illustrates the fusing of laser ultrasound data and laser ranging data (both from the LUIS system at Sacramento ALC) for a superplastically formed titanium engine seal. This data can be used for fabricating chemical milling masks that are subsequently used for precision machining of superplastically formed high performance parts.

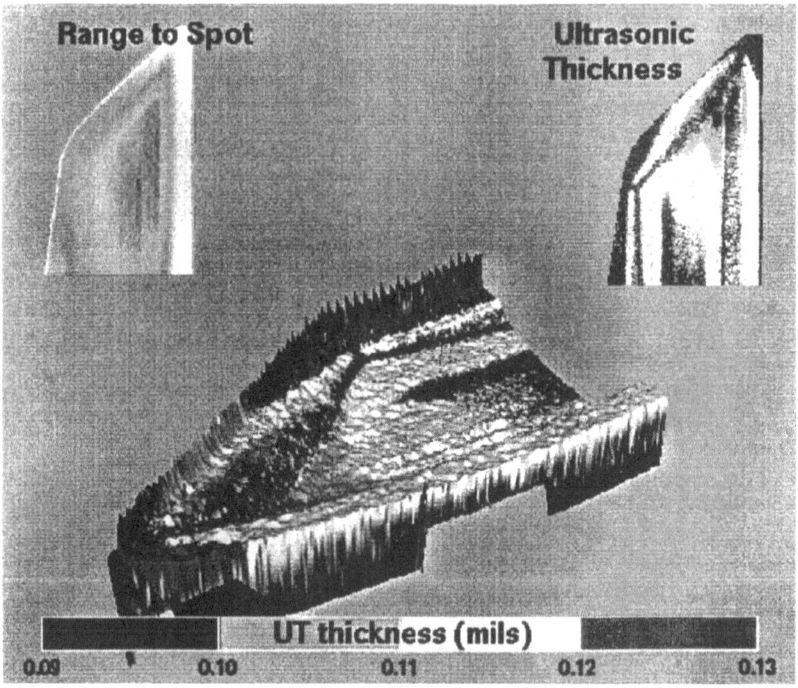

Fig. 7.9 Fusion of laser UT photogrammetric image and response data for Ti engine seal

Figure 7.10 illustrates the combination of CT images of the gauge section of a loaded adhesively bonded shear coupon. In this application, images are first co-registered in a Lagrangian sense (i.e. common centre of mass and centre of rotation), and the co-ordinates are subtracted to yield a differentiable displacement field, from which strains are calculated. This replaced a stand-alone application that had been very costly to develop without NDT data fusion tools.

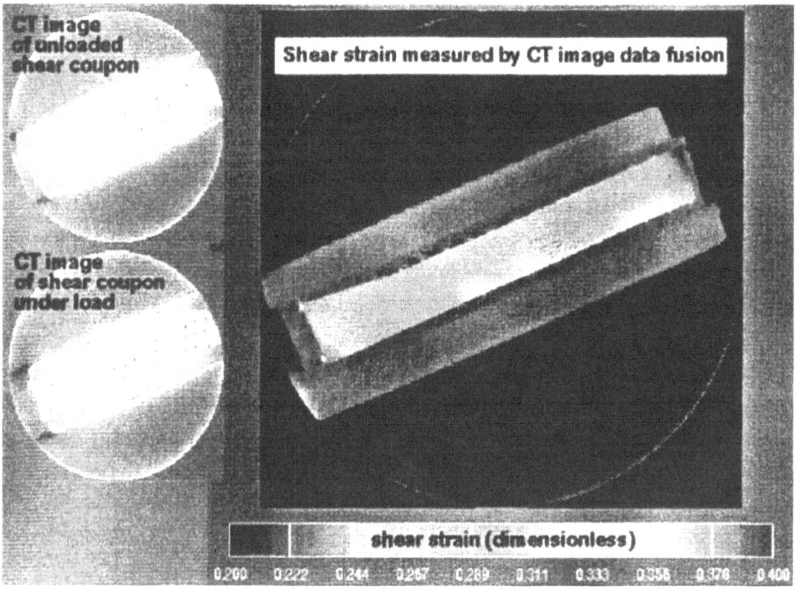

Fig. 7.10 Strain measurement by Langrangian transformation and fusion of CT images

7.6 PRESENT AND FUTURE STATE OF AEROSPACE NDE DATA FUSION APPLICATIONS

Effective NDT data fusion was first demonstrated by Bocing on thc Interim Upper Stage (IUS) rocket nozzle. Between 1985 and 1994, approximately US$3.0 millions was invested in this methodology by Boeing, the Air Force IUS program, the Air Force Rocket Propulsion Laboratory, and NASA Marshall Space Flight Center (MSFC). This methodology is credited with averting an IUS mission failure worth in excess of US$100 millions. Perceptics Corporation was also involved in the development of workstations for aerospace NDT data fusion in the 1990s, including a significant improvement to INDERS funded by NASA MSFC through the Solid Propulsion Integrity Program (SPIP). These developments in NDE data fusion for high-value SRM missions are directly applicable to other Air Force systems, as the methodology developed is very general. Emerging NDI modes such as D-Sight, infrared thermography, shearography and magneto optic provide large, information-rich data sets. The advantage of these techniques is that they can inspect relatively large areas quickly, which results in very high data rates. Because of tight schedules for the NDI/E functions at the ALCs, more efficient methods of storing, recalling, displaying and processing these data sets will become increasingly important.

The drive to affordable manufacturing is leading to an increase in the development and production of complex, monolithic structures requiring NDE data fusion functionalities. NDT data fusion technology is increasingly becoming a key requirement in these development programs and is expected to be utilised during production. NDE data fusion will be applied to post-deployment and maintenance scenarios where mission criticality is an issue. Low observable signature assurance is a particular area that falls into this category.

There are converging needs and opportunities that are continuing to accelerate the introduction of NDE data fusion technology in aerospace application. These needs and opportunities include the increasing utilisation of NDE data from more than one source to evaluate high value parts, a move to larger, more complex monolithic parts requiring unique NDE analysis and larger quantities of NDE data, the almost complete conversion to CAD methods for describing aerospace designs with a corresponding increase in demand by technical customers to superimpose the NDT data onto the CAD description of the part, and the availability of most NDT data in digital form, the relative low cost of digital processing/storage and the establishment of digital networks for rapid communication between distant sites.

REFERENCES

1. Barrett J.E., *PAM-D Star-48 failure investigation committee*, Final report, McDonnell-Douglas Document N°MDC-H1327, 1984.
2. Nelson J., Cruikshank D., Galt S., *A flexible methodology for analysis of multiple modality NDE data*, Review of Progress in QNDE, 1989, LaJolla, USA, 8A:819-826.
3. *Integrated nondestructive evaluation data reduction system (INDERS) for nozzle components*, Boeing Document N°D180-31159-1, 1988.
4. Georgeson G. E., Lempriere B.M., Shrader J.E., *Integrated nondestructive evaluation data reduction system (INDERS)*, Proc. JANNAF Rocket Nozzle Tech. Subcommittee Meeting, 1989, Silver Springs, USA, 127-136.
5. Argyle D., Youngberg J., *INDERS/2 Integrated nondestructive evaluation data reduction system*, Final report, NAS8-37801 Subtask 3.3.2.1, NASA Document N°HI-059F/1.2.9, 1994.
6. Nelson J.M., Bossi R.A., Lancaster C.A., *Nondestructive evaluation for data fusion*, Final Report, USAF Document N°AFRL-ML-WP-TR-1998-4095, 1998.
7. Poon H., Folk M., Ahmad F., *Unstructured cell data (UCD) file format*, Introduction to Postprocessing Finite Elements Results with AVS, NCSA Tutorial, University of Illinois at Urbana-Champlain, http://www.ncsa.uiuc.edu/Apps/SE/PET/courses/csm-02/lecture01/sld010.htm, 1997, slide 10.
8. Lempriere B. M., Cline J.L., *et al.*, *Ultrasonic and CT scan criteria for carbon/carbon exit cones*, Final report, IUS Special Study FSD 85-008, 1986.
9. Bossi R., Nelson J., *Data fusion for process monitoring and NDE*, Nondestructive Evaluation for Process Control in Manufacturing, Proc. SPIE, 1996, Scottsdale, USA, 2948:62-71.

8 NDT DATA FUSION FOR WELD INSPECTION

Yue Min ZHU[1], Valérie KAFTANDJIAN[2], Daniel BABOT[2]
[1]CREATIS, UMR CNRS, Villeurbanne, France
[2]CNDRI, INSA Lyon, Villeurbanne, France

8.1 INTRODUCTION

Real time, high performance and reliable inspections are desired in many fields including weld inspection, food control and medical imaging. To achieve this level of NDE, the most commonly used approach consists of applying several inspection modalities. In the field of weld inspection, ultrasonic testing (UT) is often carried out first, and radiological examination is used next to confirm the detection, identification and characterisation of defects previously revealed with ultrasounds. Sometimes, the same inspection modality is used in different operating conditions to obtain better results. For example, in conventional medical MRI, multi-spectral acquisitions are systematically performed and different images or volumes corresponding to the same organ are taken into account to achieve a more reliable and precise analysis of diseases [1-2]. At the moment, most radiological examinations in weld inspection use film radiography. This technique is slow and often does not allow 100% control. Replacing film radiography by real time X-ray imaging systems appears a reasonable alternative to reduce cost of consumables and increase inspection speed.

However, achieving the image quality of film radiography with real time apparatus remains difficult. Image quality is governed by the spatial resolution and dynamic range of X-ray imaging systems. Spatial resolution describes the contrast response of a system as a function of spatial frequency (cycles per mm or line pairs per mm). This parameter often determines the smallest defect detectable with X-ray radiography, or the sharpness (visibility) of features in a radiograph. Dynamic range is another important parameter that describes the ability of a real time X-ray imaging system to accurately measure a range of signal levels. This parameter determines directly the contrast sensitivity over the thickness variation of the material under inspection. In practical NDT applications, a low dynamic range can pose two problems if the materials or products tested present a large variation in thickness. The first is known as the blinding phenomena. To avoid this undesirable effect, one needs to collimate the material inspected. This will result in a reduction of the field of view and impossibility of controlling objects in movement. The second problem is over-exposure or under-exposure of image regions. Limited dynamic range reduces significantly contrast sensitivity and consequently the probability of detection.

To improve the spatial resolution of real time X-ray imaging systems, two types of approaches have been reported in the literature. The first is a hardware-based approach, which uses smaller CCD cells and multiple line CCD techniques [3]. The other is a software-based approach, in which image restoration techniques have been developed to obtain high spatial resolutions beyond the capability of X-ray detectors [4-5]. To improve the dynamic range of X-ray imaging systems, the same approaches are followed. In general, achieving a very high spatial resolution or wide dynamic range is technically very difficult and costly. It is also limited because sensitive cells should have a size sufficiently large to capture X-ray energy. Additionally, the lowest levels are limited by electronic noise and the highest levels by detector saturation. Therefore, software development appears as an interesting solution.

Meanwhile, the synergistic use of an improved filmless X-ray imaging system and an ultrasonic C-scan make the achievement of lower exploitation/production costs and higher inspection reliability possible.

This chapter contains a description of software techniques used for improving the dynamic range of X-ray imaging systems, and thus increasing weld inspection reliability. These techniques use a data fusion approach based on Dempster-Shafer (DS) theory. They exploit redundant and complementary information either from the same inspection modality (e.g. in the case of dynamic range improvement) or from different inspection modalities.

8.2 ILLUSTRATION OF THE PROBLEM

This section introduces two NDE problems and shows that fusing data from two sources can solve them.

Figure 8.1 shows two images of the same step-plate aluminium phantom which presents a thickness range of 90 mm (9 steps of 10 mm in thickness). These were acquired with a linear solid-state detector based X-ray imaging system, which presents in general a wider dynamic range than the popular X-ray image intensifier (XRII) based imaging system [6]. These were obtained at two different X-ray energies and can be referred as strong and weak X-ray images (100 kV, 22mA and 70kV, 20mA, respectively).

Acquiring two images for the same step-plate was necessary since the nine steps were not visible in a single image. This is because the dynamic range of the X-ray imaging system is too small with respect to the variation in thickness of the object. It can be observed that all the steps of the phantom have been made visible in neither of the two images. In the weak X-ray image (right in figure 8.1), thicker steps appear dark and confused due to under-exposure, and in the strong X-ray image (left in figure 8.1), thinner steps are saturated due to over-exposure. The goal is to bring out all the steps present in the step-plate as if it has been imaged with a perfect and unlimited dynamic range radiographic system. Compensating for the insufficient dynamic range can be considered as a problem of fusing multiple images such that all details present in the object are visible in the fused image.

Fig. 8.1 Illustration of the effect of limitation of the dynamic range of a X-ray imaging system with a step-plate aluminium phantom of nine steps

In the above-illustrated problem, the two sources to be fused correspond to two different acquisitions with the same imaging modality. In the following problem concerning weld inspection, data fusion techniques are used for two different inspection modalities. Indeed, in practical weld inspection applications, UT is often used first and followed by film radiography to confirm the UT results. That is because UT offers very high sensitive detection of defects, even when the defects have small thickness and allows an estimation of defect depth. However, with ultrasonic inspection, identification and sizing of defects are sometimes difficult. For that reason, film radiography is used to obtain directly the morphology of defects. But film radiography fails in detecting defects presenting small thickness, and does not give any information about the depth of defects. Therefore, film radiography and UT are two complementary inspection modalities. To exploit complementary information from these two inspection modalities, data fusion techniques have to be developed.

The practical design and implementation of a fusion system for a given application involve several aspects such as information modelling, data analysis and processing, data combination and interpretation of fused results. They are addressed next for different applications.

8.3 INFORMATION FUSION FROM TWO SOURCES USING DEMPSTER-SHAFER THEORY

Data fusion is a technique allowing to take simultaneously into account heterogeneous data from different sources in order to make optimal decisions. It is a relatively new information processing technique in the field of NDT. It is also currently an active research area for the computer vision and image processing community. The idea of performing data fusion stems from the fact that information from one single data source is often incomplete, uncertain and imprecise; but the combined use of complementary (and redundant) information from different data sources makes it possible to extract more pertinent and desired information. Numerous data fusion applications have been reported that deal with process control [7], target tracking [8], robot navigation [9], object recognition [10], medical image analysis [11-12], NDT [13-15]. Details of the state of the art in the field of data fusion can be found in the literature [16-17].

Regardless of its application, the goal of data fusion is to reduce uncertainty, imprecision, inconsistency, or incompleteness by combining both redundant and complementary data. A few mathematical formalisms (e.g. probability, fuzzy logic, possibilities, evidence theory) are available to perform a measure of the uncertainty and imprecision. Among these, evidence theory, also called DS evidence theory [18], is a powerful and flexible mathematical tool for handling uncertain and inaccurate information. First, by representing the uncertainty and the imprecision of a body of knowledge via the notion of evidence, belief can be committed to one single hypothesis (singleton) or a composite hypothesis (union of hypotheses). Next, the evidence combination rule of the DS theory provides an interesting operator to integrate multiple information from different sources. Finally, decision on the optimal hypothesis choice can be made in a flexible and rational manner.

Dempster-Shafer evidence theory was developed as an attempt to generalise probability theory. This theory is suitable to reason with uncertainty and has been developed to overcome the limitation of conventional probability theory by distinguishing between uncertainty and imprecision. This is achieved in particular by making it possible to handle composite hypotheses. It is also suited for combining information from different sources. In DS theory, there is a fixed set of N mutually exclusive and exhaustive elements, called the frame of discernment, symbolised by $\Theta = \{H_1, H_2, \ldots, H_N\}$. The representation scheme Θ defines the working space for the desired application since it consists of all propositions for which the information sources can provide evidence. Information sources can distribute mass values on subsets of the frame of discernment, $A_i \in 2^{\Theta}$. An information source assigns mass values only to those hypotheses, for which it has direct evidence. That is, if an information source can not distinguish between two propositions A_i and A_j, it assigns a mass value to the set including both propositions $(A_i \cup A_j)$.

The derivation of the mass distribution is the most crucial step since it represents the knowledge about the actual application as well as the uncertainty incorporated in the selected information source:

$$0 \le m(A_i) \le 1 \tag{8.1}$$

$m(A_i) > 0$, A_i is called a focal element. The mass distribution for all the hypotheses has to fulfil the following conditions:

$$m(\varnothing) = 0$$
$$\sum_{A_i \in 2^{\Theta}} m(A_i) = 1 \tag{8.2}$$

The mass distributions m_1 and m_2 from two different information sources are combined using Dempster's orthogonal rule. The result is a new distribution, $m = m_1 \oplus m_2$, which carries the joint information provided by the two sources:

$$m(A_i) = (1-K)^{-1} \times \sum_{Ap \cap Aq = A_i} m_1(A_p) m_2(A_q) \tag{8.3}$$

where

$$K = \sum_{Ap \cap Aq = \phi} m_1(Ap) m_2(Aq)$$

K is often interpreted as a measure of conflict between the different sources and it is introduced in equation (8.2) as a normalisation factor. The larger K, the higher the conflict between sources, and the lower the meaning of their combination.

From a mass distribution, two functions can be evaluated that characterise the uncertainty about the hypothesis A_i. The belief function *Bel* of A_i is obtained by summing all masses supporting A_i, measures the minimum uncertainty value about A_i whereas plausibility *Pls* reflects the maximum uncertainty value about this hypothesis. These two measures span an uncertainty interval $[Bel(A_i), Pls(A_i)]$ called "belief interval". The length of this interval gives a measure of imprecision about the uncertainty value. Belief and plausibility functions are defined from 2^Θ to $[0,1]$:

$$\mathrm{Bel}(A_i) = \sum_{A_j \subseteq A_i} m(A_j) \tag{8.4}$$

$$\mathrm{Pls}(A_i) = \sum_{A_j \cap A_i \neq \phi} m(A_j) \tag{8.5}$$

These measures, which have been sometimes referred to as lower and upper probability functions, have the following properties:

$$Pls(A_i) = 1 - Bel(\overline{A_i}) \tag{8.6}$$

$$\mathrm{Bel}(A_i) \leq \mathrm{Pls}(A_i) \tag{8.7}$$

where $\overline{A_i}$ is the complementary of A_i : $A_i \cup \overline{A_i} = \Theta$ and $A_i \cap \overline{A_i} = \phi$

In data fusion, belief and plausibility functions are calculated with the mass functions resulting from Dempster's combination. Decision about the choice of hypotheses is made using one of the criteria of maximum plausibility and maximum belief. It is important to note that the two decision criteria do not always lead to the same results. The maximum belief corresponds to optimistic approaches and the maximum plausibility to prudent approaches. The following example illustrates how the DS theory is used to solve the problem of weld defect detection with radiography and ultrasonic. Let's define the following three hypotheses:

- H_1 : no defect,
- H_2 : linear defects (e.g. lack of fusion/penetration, cracks),
- H_3 : porosity.

The frame of discernment is $\Theta = \{H_1, H_2, H_3\}$. The composite hypothesis $H_2H_3 = H_2 \cup H_3$ means the presence of defects which can be linear defect or porosity, and Θ the total ignorance about the hypotheses. If the X-ray and ultrasonic inspections give the following mass distributions, respectively: $m_{rx}(H_3)=0.6$, $m_{rx}(\Theta)=0.4$, $m_{ut}(H_2)=0.95$, $m_{ut}(\Theta)=0.05$, the combined mass can be obtained using equation (8.3), as illustrated in table 8.1. It can be written:

$$
\begin{aligned}
K &= m_{rx}(\text{porosity})\, m_{ut}(\text{linear defects}) &&= 0.6 \times 0.95 &&= 0.57 \\
m(\text{linear defects}) &= m_{ut}((\text{linear defects})\, m_{rx}(\Theta)/(1\text{-}K) &&= 0.95 \times 0.4/0.43 &&= 0.884 \\
m(\text{porosity}) &= m_{rx}(\text{porosity})\, m_{ut}(\Theta)/(1\text{-}K) &&= 0.6 \times 0.05/0.43 &&= 0.070 \\
m(\Theta) &= m_{ut}(\Theta)\, m_{rx}(\Theta)/(1\text{-}K) &&= 0.05 \times 0.4/0.43 &&= 0.047
\end{aligned}
$$

Following the DS combination, the quantum of evidence committed to singletons is reduced, but this reduction is much more important for porosity hypothesis than for linear defects hypothesis. At the same time, ignorance is also reduced in particular with respect to X-ray inspection. Since, no composite hypotheses are involved in this example, mass, belief and plausibility functions are equal. As a result, it is the hypothesis H_2 that would be chosen after fusing X-ray and ultrasonic data, provided the degree of conflict K is acceptable. For the same example, if the two inspection modalities give the following evidence, for the same frame of discernment:

$$
\begin{aligned}
m_{rx}(H_3) &= 0.8 \qquad & m_{rx}(\Theta) &= 0.2 \\
m_{ut}(H_2H_3) &= 0.8 \qquad & m_{ut}(\Theta) &= 0.2
\end{aligned}
$$

we will obtain (table 8.2):

$$
\begin{aligned}
m(\text{defects}) &= m_{rx}(\Theta)\, m_{ut}(\text{defects}) = 0.2 \times 0.8 = 0.16 \\
m(\text{porosity}) &= m_{rx}(\text{porosity})\, m_{ut}(\text{defects}) + m_{rx}(\text{porosity})\, m_{ut}(\Theta) \\
&= (0.8 \times 0.8) + (0.8 \times 0.2) = 0.8 \\
m(\Theta) &= m_{rx}(\Theta)\, m_{ut}(\Theta) = 0.2 \times 0.2 = 0.04
\end{aligned}
$$

Thus, after DS combination, evidence committed to the hypothesis porosity hasn't changed, but that to the composite hypothesis {linear defects or porosity} as well as to ignorance has been largely reduced. This indicates that imprecision and uncertainty have been significantly reduced. From the above mass distributions, belief and plausibility functions are calculated as:

$$
\begin{aligned}
Bel(\text{porosity}) &= m(\text{porosity}) = 0.8 \\
Bel(\text{defects}) &= m(\text{porosity}) + m(\text{defects}) = 0.96 \\
Pls(\text{porosity}) &= m(\text{porosity}) + m(\text{defects}) + m(\Theta) = 1.0 \\
Pls(\text{defects}) &= m(\text{porosity}) + m(\text{defects}) + m(\Theta) = 1.0
\end{aligned}
$$

Table 8.1 Computation of the combined mass distribution (case 1)

m_{ut} \ m_{rx}	H_3	Θ
H_2	ϕ	H_2
Θ	H_3	Θ

Table 8.2 Computation of the combined mass distribution (case 2)

m_{ut} \ m_{rx}	H_3	Θ
H_2H_3	H_3	H_2H_3
Θ	H_3	Θ

8.4 IMPROVEMENT OF DYNAMIC RANGE

It is now well known that image restoration techniques can be used to obtain high spatial resolutions beyond the capability of a X-ray detector [4-5]. This has made the use of X-ray imaging more efficient for NDT applications, and has also expanded its application field. Following the same idea, it is possible to improve the dynamic range of X-ray imaging systems using data fusion techniques. The idea of improving the dynamic range by data fusion consists of integrating information from the same real time X-ray imaging system, but obtained under two different acquisition conditions. Each acquisition provides an accurate measurement of signal levels over a certain range. The fusion of the signal level measurements over two different ranges results in a significant improvement of the final dynamic range. Section 8.4.1 presents a data fusion method based on DS theory. In section 8.4.2, the possibility of using this method for improving dynamic range is demonstrated with the aid of simulated images. Finally, application examples on real specimens and under real acquisition conditions are shown in section 8.4.3.

8.4.1 Data Fusion Scheme Based on Dempster-Shafer Theory

The problem of improving dynamic range of a real time X-ray imaging system with data fusion can be defined as the fusion of two images from the same object into a single image. These two images have been acquired with the same imaging system operating in two different modes. In terms of DS approach this means determining the object by integrating images obtained under two different acquisition conditions. The fusion approach based on the DS theory consists of four steps:

- an initial number of classes (or hypotheses) is defined,
- histogram models are constructed for the images to be fused, and mass functions are obtained from the grey level histogram models,
- pixels from the images are fused by combining their mass functions using Dempster's orthogonal rule,
- decision is made using one of the criteria of maximum belief and maximum plausibility, which are all calculated with the fused mass functions.

Determining mass functions from image histogram is illustrated in figure 8.2. For each grey level of each image, a histogram is modelled by Gaussian functions, from which mass functions are then derived. For a given hypothesis H_i, the probability of having the grey level x_j, is defined as:

$$P(x_j / H_i) = \frac{1}{\sigma_i \sqrt{2\pi}} \exp^{-\frac{(x_j - \overline{H}_i)^2}{2\sigma_i^2}} \tag{8.8}$$

where $\overline{H}_i$ and σ_i are respectively the mean and the standard deviation of the Gaussian expression.

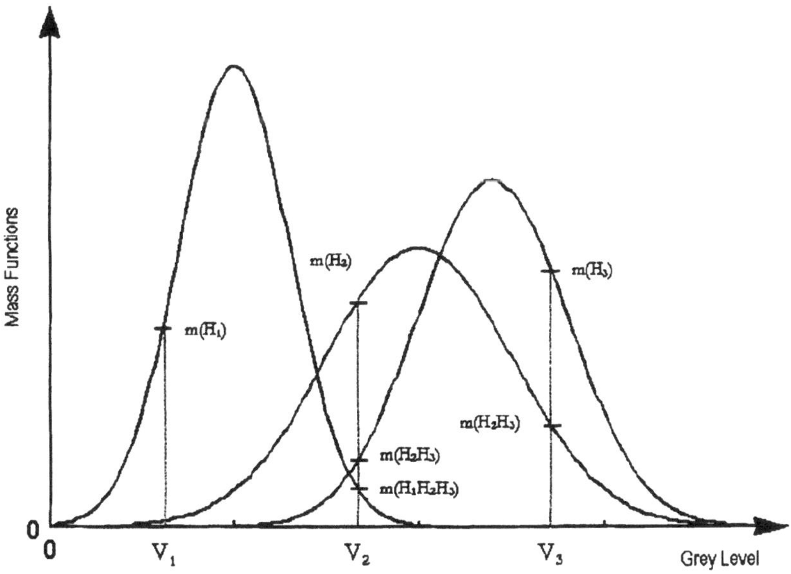

Fig. 8.2 Modelling of mass functions from image histograms

Let $\{H_1, H_2, ..., H_N\}$ be a set of simple hypotheses (e.g. objects or regions in the fused image) representing the frame of discernment. One can see in figure 8.2 that, for certain grey levels, only one class has significantly higher probability, which is the case for the grey level V_1. This corresponds to the case of one focal element that is formed of one simple hypothesis. The mass function associated with this simple hypothesis is defined as:

$$m_j(H_i) = P(x_j/H_i) \tag{8.9}$$

In other cases, several classes have significant probability values, which is the case for the grey levels V_2 and V_3. In this situation, different hypotheses are ranged in decreasing order of probability, and consonant masses are generated that correspond to the focal elements nested like $A_1 \subset A_2 \subset \dots A_k$. The first element is a simple hypothesis that has the greatest probability value. Its associated mass corresponds to the length of a segment between it and its following hypothesis. The other elements are composite hypotheses. The mass functions for consonant elements are mathematically expressed next.
If $P(x_j/H_a) > P(x_j/H_b) > P(x_j/H_c)$, then:

$$\begin{aligned} m_j(H_a) &= P(x_j/H_a) - P(x_j/H_b) \\ m_j(H_aH_b) &= P(x_j/H_b) - P(x_j/H_c) \\ m_j(H_aH_bH_c) &= P(x_j/H_c) \end{aligned} \tag{8.10}$$

If no class has significant probability, the mass will be committed to the total ignorance $m(\Theta)=1$. Once mass functions are calculated for all grey levels, normalisation is performed using the maximal value of the histogram. This is necessary in order to satisfy equation (8.2). After determining the mass functions, the next step is to combine them using Dempster's rule of combination (equation (8.3)). Assigning the grey level to a class is performed on the combined mass functions. This decision is made using one of the criteria of maximum plausibility and maximum belief.

The fusion of images is an iterative process involving each time the steps of mass determination, mass combination and decision. The process is reiterated because, after the first iteration the statistics of each class of pixels (e.g. population, mean, standard deviation) have changed. The new values of each class are recalculated and used as prior knowledge for the next iteration. At this stage, different tests are also performed to adapt the number of classes to the present situation. A class which becomes empty is removed. If a class becomes too heterogeneous in terms of standard deviation and population, it is split into two different separated subsets. The third case is that if the means of the two classes are close in the two images, they are fused into one single class. Starting from adapted initial conditions, the classification process is iterated until the algorithm converges, that is until pixel classes no longer change from one iteration to the next.

8.4.2 Simulations

The dynamic range is often defined as the ratio between the saturation signal and noise. A practical method for evaluating the overall and/or useful dynamic range performance of a real time X-ray imaging system is to use a step-plate phantom which presents a number of steps, each step corresponding to a given thickness [6]. Evaluation of dynamic range parameter is carried out by determining the number of steps visible in the final digital image of the step-plate phantom. Obviously, a wide dynamic range system should make it possible to distinguish a large number of steps having a thickness variation over a wide range. In most cases, since real time X-ray imaging systems present limited dynamic range, only a certain number of steps can be visible in the image; other steps having completely disappeared because of over-exposure or under-exposure.

In order to simulate two acquisition conditions, two images have been created to represent the X-ray images of a step-plate phantom acquired under two operational conditions (i.e. voltage and current of the X-ray tube). The first image simulates strong X-rays (i.e. higher voltage and current). That is, thinner steps are not very well defined because of the detector saturation. In contrast, thicker steps are well separated, but noise, which is assumed to be Gaussian, is generally more important. The second image represents weak X-rays (i.e. lower voltage and current). In this image, thicker steps are under-exposed, and appear commonly dark on the image. Thinner steps are well detected and distinguishable from each other. Moreover, noise is generally less important with weak X-rays than with strong ones. In both cases, images are characterised by the presence of different rectangular regions, which correspond to the different thickness steps.

To fuse the two images using the previously described fusion approach, the grey levels are considered basic properties of a pixel. The different hypotheses are the steps of the phantom. Each pixel pair represents exactly the same physical point. So, a registration pre-processing, necessary in most practical applications, can be omitted for the present simulation case.

The two simulated images of size 256 × 256 pixels are shown in figure 8.3. The three rectangular step regions (corresponding to three classes) an equal area, each of them occupying 10 % of the total surface. The background represents 70% of the total surface. The images are corrupted by an additive Gaussian noise. To simulate acquisition conditions, the means of the four classes are not equally spread over the grey level scale. For the strong X-ray image (left image in figure 8.3), the means of the background and the thinnest region are close. For the weak X-ray image (right image in figure 8.3), the means of the two thickest step regions are close. It is therefore impossible to distinguish the four classes from the histograms of each image, and the global histogram seems to be tri-modal. The goal of the fusion operation in this example is to determine the thickness of each step and to classify the pixels into three steps plus the background.

Figure 8.4 shows the fusion results corresponding to three iterations (1, 3 and 9). The algorithm has converged at iteration 9. The decision criterion used here is the maximum belief on a simple hypothesis. For the same initial images and the same iterations, the fusion results obtained with the criterion of maximum plausibility are shown in figure 8.5. It is seen that the results obtained with these two criteria are very similar.

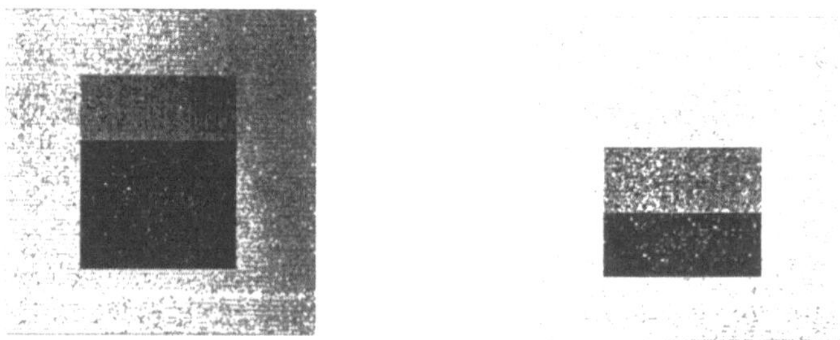

Fig. 8.3 Simulated images used for evaluating data fusion algorithm

Fig. 8.4 Fusion results obtained using the maximum belief on a simple hypothesis (left: iteration 1, middle: iteration 3, right: iteration 9 and convergence)

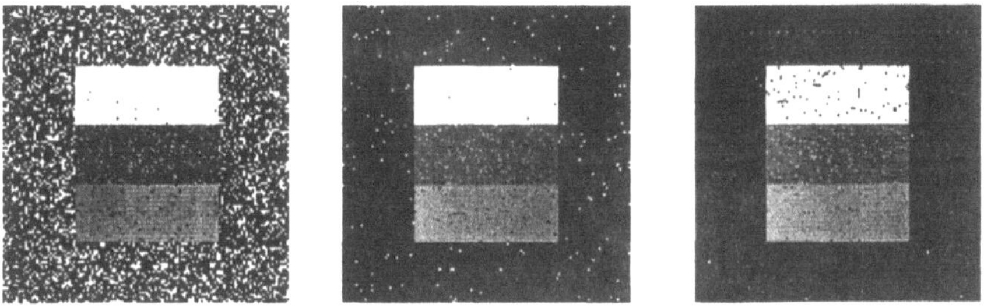

Fig. 8.5 Fusion results obtained using the maximum plausibility on a simple hypothesis (left: iteration 1, middle: iteration 3, right: iteration 9 and convergence)

8.4.3 Case Studies

Figure 8.6 shows the same images (top and middle in figure 8.6) as those in figure 8.1. As previously indicated, they were obtained with to two different acquisition conditions. In these images, a horizontally oriented linear defect in the step-plate phantom is visible.

Unlike in the simulation cases, real images should be registered before performing data fusion. Indeed, geometrically, each pair of pixels in the two images does not necessarily represent the same physical point in the object. The registration method used in the present study was based on maximisation of the correlation of two images. Once the registration has been performed, the fusion algorithm was computed in the same manner as in the simulation case. The fused image is shown on the bottom of figure 8.6. If a pixel is assigned to a simple hypothesis, the grey level was selected so that it corresponds to a physical step. If the pixel is assigned to a composite (double) hypothesis, an intermediate grey level was chosen. In the fused image, all the steps of the phantom have been brought out and separated. It can be considered as a radiograph obtained with a X-ray imaging system having ideal dynamic ranges. It should be underlined that the above labelling is just a problem of visualisation of the fused results. The fused regions can be labelled in any other manners. The linear defect in the phantom has also been detected as double hypothesis. That is, no new class has been created for this defect. It was considered by the algorithm as a result of the union of the two closest steps. In general, few pixels have been classified into double hypothesis. In most cases, isolated pixels representing noise and pixels situated on borders between two classes are likely to assign to a double hypothesis.

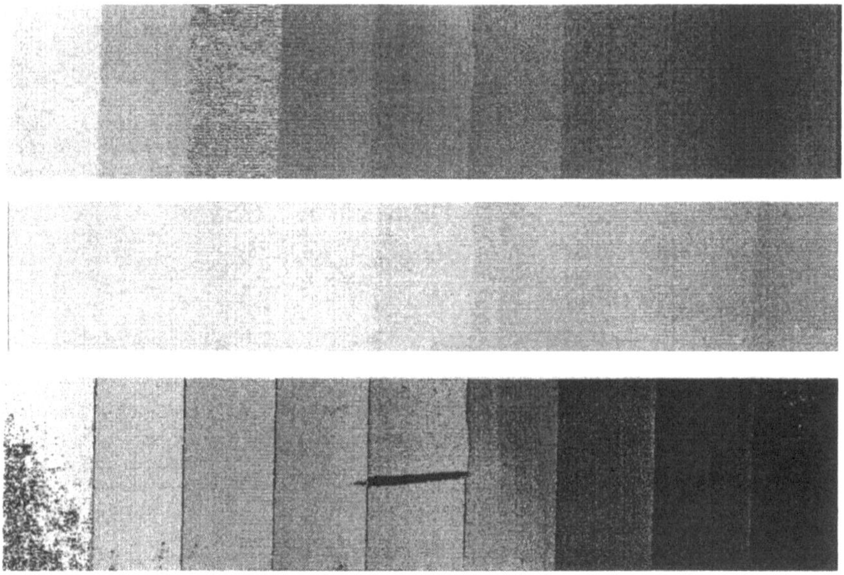

Fig. 8.6 Improvement of dynamic ranges in X-ray imaging by fusing images acquired with two operating conditions (top: weak X-ray image, middle: strong X-ray image, bottom: fused image)

8.5 WELD INSPECTION

For the application of weld inspection involving ultrasonic and X-ray imaging, the DS theory of evidence presents several interesting aspects. In cases where knowledge about a defect is available only for one of the two imaging modalities, the masses can be assigned only to the defect seen by this imaging modality, and the ignorance is assigned to the other imaging modality. When a defect is detected with ultrasound but not with radiography, the DS theory allows the simple hypothesis to be associated with the ultrasonic source, and the composite hypothesis with the radiographic source. In cases where both imaging modalities are not able to distinguish between two types of defects, we are not obliged to assign masses to these two defects separately, but the masses will be assigned to the union of the defects (composite hypothesis). Other modalities can also be used for inspecting weld quality, an example of which is eddy current and ultrasonic inspections.

In the synergistic use of X-rays and ultrasounds, the ultrasonic modality is often exploited under the form of a 2-D image or one-dimensional signal, whereas a radiograph is always used as a 2-D image. Thus, the fusion of both images will be performed as well as the fusion of the ultrasonic one-dimensional signal and radiograph.

In all application problems using the DS theory, the appropriate determination of the mass function is a delicate point. In its early use for approximating reasoning with artificial intelligence, the mass functions were often obtained in a heuristic or empirical manner [19-21]. The fusion was also often done at high level (e.g. fusion of objects). The use of heuristic values could provide reliable results only for a particular configuration of the application in question, but little reproducibility. The exploitation of histograms provides, as presented in section 1.4, a useful way for automatically determining mass functions.

For the present application of weld inspection, a data fusion method based on the combined use of the DS theory and fuzzy logic is described. Fuzzy sets were introduced in 1965 by Zadeh as a generalisation of conventional set theory [22]. The fuzzy set theory uses vagueness for formulating and solving various problems. Conventional sets contain objects that satisfy precise properties required for membership, while fuzzy sets contain objects that satisfy imprecise properties of varying degrees. For example, let's consider the set of numbers F that are close to the number seven. Zadeh proposed representing F by a membership function noted μ_F that maps numbers into the entire unit interval [0,1]. The membership function is the basic idea in fuzzy set theory; its values measure degrees to which objects satisfy imprecisely defined properties. The approach is suitable for both low and high level fusion. In the case of low level fusion (e.g. pixel-by-pixel fusion), the fuzzy approach involved concerns the fuzzy C-means (FCM) clustering. In the case of high level fusion (e.g. fusion of processed data), fuzzy inference is used to determine the assignment of masses to the hypotheses.

8.5.1 Pixel Level Fusion of Radiographic and Ultrasonic Images

The idea of using in a combined manner FCM and DS theory to fuse radiographic and ultrasonic images at the pixel level is explained next. The fusion of two pixels is achieved through integrating mass functions associated with the grey levels corresponding to the two pixels. The principle is to assign a mass function value to each pixel grey level using the FCM clustering. The latter is used to represent the grey levels as fuzzy sets. Each pixel is characterised by a degree of membership to a given cluster, or class, or still hypothesis (these three terms will be used in an indifferent manner in the text). The types of classes to be considered (i.e. simple or composite classes) as well as the corresponding mass function values are derived from the membership functions. Once the mass functions are determined for each image, the DS combination rule and decision are applied to fuse them.

In this DS theory based segmentation paradigm, it is evident that the appropriate determination of mass functions plays a crucial role since classification of a pixel to a cluster is directly yielded from the mass functions. Moreover, the mass functions take into consideration pixels' properties from a viewpoint of fuzziness. That is rather different from the approach presented in section 8.4, and also different from the many DS fusion applications, in which the determination of mass functions is not driven by image data. In all cases, the determination of mass functions can introduce subjective factors in the fusion process. It is thus necessary to reduce this subjectivity to a minimum. A salient aspect of FCM clustering is its unsupervised character. Therefore, determining mass functions from FCM is completely data-driven. Generally, the boundaries of real data structures (clusters) are not well defined (fuzzy) and the transition from one cluster to another is not abrupt but smooth. This means that pixel information comes from several clusters. As pointed out by Li and Yang, image processing bears some fuzziness in nature [23]. Operations for classifying a pixel to clusters are also subjective. There are intrinsic ambiguity and vagueness in the clustering process. To cope with this situation, a fuzzy approach such as FCM is an interesting pixel clustering model. It has the capability to handle ambiguity and vagueness attached to pixel grey level value which come from the imperfections of imaging systems. Indeed, compared to conventional mathematics, there are no well-defined boundaries in fuzzy logic, and the transition between full membership and no membership is gradual. Also, fuzzy segmentation relies on the fact that each pattern is characterised by a degree of membership to a given cluster. This enables us to distinguish between patterns forming a main structure of the cluster and those being outliners [24].

Fuzzy Clustering

In the case of a general formulation, we can denote any data to be classified by a vector $X = \{x_1, \ldots, x_M\}$, which represents a finite data set of dimension M. Let $C \geq 2$ be an integer designating the number of clusters in which X will be classified, and $\Re^{C\times M}$ the set of all real $C \times M$ matrices. A fuzzy C-partition of X is represented by a matrix $U = [\mu_{ik}] \in \Re^{C\times M}$, U is used to describe the clusters of X, and where $\mu_{ik} = \mu_i(x_k)$ expresses the degree of membership of the element x_k to cluster i, and verifies the following constraints:

$$\mu_{ik} \in [0,1] \quad 1 \le i \le C; \quad 1 \le k \le M$$
$$\sum_{i=1}^{C} \mu_{ik} = 1 \quad 1 \le k \le M \tag{8.11}$$
$$\sum_{k=1}^{M} \mu_{ik} > 0 \quad 1 \le i \le C$$

In the context of image segmentation, the FCM can be implemented to work on one-dimensional attribute such as the grey-level. Let *HS* be the histogram of image of *L*-levels, where *L* is the number of grey levels. Each pixel has a feature that lies in the discrete set $X = \{0, 1,..., L\text{-}1\}$. In this case, each element of the data set represents a grey level value, and $\mu_{il} = \mu_i(l)$ can be used to express the degree of membership of the element l to cluster i. A good partition *U* of *X* is then yielded by the minimisation of the FCM [25]:

$$J_q(U, V : L) = \sum_{l=0}^{L-1} \sum_{i=1}^{C} (\mu_{il})^q \cdot HS(l) \cdot \| l - v_i \|_A^2 \tag{8.12}$$

where $q \in [1, +\infty[$ is a weighting exponent called the fuzzifier, and $V = (v_1, v_2,..., v_C)$ the vector of the cluster centres. $\|x\|_A = \sqrt{x^T A x}$ is any inner product norm where *A* is any positive definite matrix. The FCM clustering consists of the following steps:

- fix the number of clusters *C* for $2 \le C \le L$, and the threshold value ε,
- find the number of occurrences, $HS(l)$, of the level *l* for $l = 0, 1, 2, ..., L\text{-}1$;
- initialise the membership degrees μ_{il} using the *L* grey levels such that:

$$\sum_{i=1}^{C} \mu_{il} = 1 \quad l = 0, 1, ..., L-1$$

- compute the centroid v_i as follows :

$$v_i = \frac{\sum_{l=0}^{L-1} (\mu_{il})^q \cdot HS(l) \cdot l}{\sum_{l=0}^{L-1} (\mu_{il})^q \cdot HS(l) \cdot l} \quad i = 1,...,C \tag{8.13}$$

- update the membership degrees:

$$\tilde{\mu}_{il} = \left[\sum_{j=1}^{C} \left(\frac{\left\| l - v_i \right\|_A}{\left\| l - v_j \right\|_A} \right)^{\frac{2}{(q-1)}} \right]^{-1} \tag{8.14}$$

- compute the defect measure:

$$E = \sum_{i=1}^{C} \sum_{l=0}^{L-1} | \tilde{\mu}_{il} - \mu_{il} |$$

$$\text{if } (E > \varepsilon) \; \{\mu_{il} \leftarrow \tilde{\mu}_{il} \text{ go to (d)}\}$$

- defuzzification process.

The FCM algorithm alternates between equations (8.13) and (8.14), and converges iteratively to either a local minimum or a saddle point of J_q.

Mass Functions Determination Using FCM

Let I_1 and I_2 be the two images to be fused. As before, since fusion operation is performed at the pixel level, to correctly fuse I_1 and I_2 a precise geometrical correspondence should exist between the two images to ensure that each pair of pixels in the two images does represent the same physical point in the object. Therefore, the images under consideration are assumed to match perfectly. Otherwise, images have to be registered prior to fusion.

The first step in computing mass functions using FCM is to represent each grey level l of the images with a function of membership μ_{il} to class i. To do that, FCM is applied to I_1 and I_2 . Then, pieces of evidence in DS theory are represented by membership functions. The hypotheses being considered in Dempster-Shafer formulation are the following: ϕ (whose mass is null), simple hypotheses H_i and composite hypotheses $H_r \cup \ldots \cup H_s$. In the present study, we consider only double hypotheses $H_r \cup H_s$ as composite hypotheses. This is because when performing clustering using solely grey level information, the intersection occurs only between two curves $\mu_r(l)$ and $\mu_s(l)$ corresponding to two consecutive centroids v_s and v_r. The intersection point corresponds to the highest degree of ambiguity between hypotheses H_r and H_s. Another argument resides in the fact that, grey levels presenting a high level of ambiguity in their memberships correspond to pixels located at the edges of two clusters, such as r and s, on the image. Thus, for these pixels the ambiguity is dealt with two neighboured clusters. A double hypothesis is created according to the following strategy:

- Assign a non-null mass to $H_r \cup H_s$ if H_r and H_s are not discriminated in the images (i.e. not distinguishable by sensors). This corresponds to a case in which there is an ambiguity between H_r and H_s.
- Assign a null mass to $H_r \cup H_s$ if H_r and H_s are discriminated in the image. This reflects the situation where there is no or little membership of l to both clusters r and s simultaneously.

Therefore, a non-normalised mass $\tilde{m}$ of hypothesis A_i is evaluated depending on the simple or double type hypothesis. The final mass is obtained after normalisation with respect to the entire set 2^Θ, that is:

$$m(A_i) = \frac{\tilde{m}(A_i)}{\sum_{A_j \in 2^\Theta} \tilde{m}(A_j)} \tag{8.15}$$

such that:

$$\sum_{A_j \in 2^\Theta} m(A_j) = 1 \tag{8.16}$$

The non-normalised mass functions $\tilde{m}$ are computed from the membership functions $\mu_i^{m,n}$ following two cases;

Simple Hypothesis

Masses of simple hypotheses H_i are directly obtained from the filtered membership functions $\mu_i^{m,n}$ of the grey level $l(m,n)$ to cluster i as follows:

$$\tilde{m}(A_i) = \tilde{m}(H_i) = \mu_i^{m,n}, \quad 1 \le i \le N \tag{8.17}$$

Double hypothesis

If there is ambiguity in assigning a grey level $l(m,n)$ to cluster r or s, that is $|\mu_r(l) - \mu_s(l)| < \xi$ where ξ is a thresholding value, then a double hypothesis is formed and its associated mass is calculated according to the following formula:

$$\tilde{m}(A_i) = \tilde{m}(H_r \cup H_s) = \Im(\mu_r(l), \mu_s(l)) \tag{8.18}$$

In equation (8.18), $\mathfrak{I}$ is a function generating the mass attributed to the union of clusters r and s using the membership degrees of l to these two clusters. It is evident that the more $\mu_r(l)$ or $\mu_s(l)$ is important, the more the mass to their union. At the same time, the closer the values of $\mu_r(l)$ and $\mu_s(l)$ are, the greater the ambiguity between the two hypotheses forming their union, and consequently the greater the mass to assign to the double hypothesis $H_r \cup H_s$. Therefore, to determine the function $\mathfrak{I}$, two constraints should be taken into account:

- for a given grey level l, the mass assigned to the double hypothesis $H_r \cup H_s$ should be proportional to the membership degrees of l(m,n) to each of hypotheses H_r and H_s;
- the mass assigned to $H_r \cup H_s$ should be inversely proportional to the (absolute) difference between the membership degrees of l(m,n) to hypotheses H_r and H_s.

The problem can be formulated by considering $\mathfrak{I}$ as a new variable, which represents the surface of a triangle. The surface of such a triangle will depend both on the amplitudes of the membership functions of l to clusters r and s, and on the difference between these amplitudes. Figure 8.7 shows how the triangle is constructed and how the mass of the double hypothesis $H_r \cup H_s$ is derived from the surface of the triangle. The vertical axis of figure 1.7 represents the membership degrees. The unit of the horizontal one has no importance on the mass computation. In practical implementations, the horizontal axis is considered normalised, and therefore has no unit. The two dotted triangles represent two triangular membership functions corresponding to classes r and s, between which ambiguity is to be studied. The two triangles are isosceles, and have the same length for their bases, which is equal to twice the distance between the vertexes of the two triangles. The heights of the triangles are equal to $\mu_r(l)$ and $\mu_s(l)$, respectively. The overlapping surface, as represented by the shaded region S in figure 8.7, of the two triangles represents the membership of a grey level to the composite hypothesis $H_r \cup H_s$ in the new reference system. It can be regarded as the level of ambiguity associated with the classification of a pixel to one or the other cluster. Therefore, the mass value attributed to the double hypothesis $H_r \cup H_s$ can be directly calculated from the surface S.

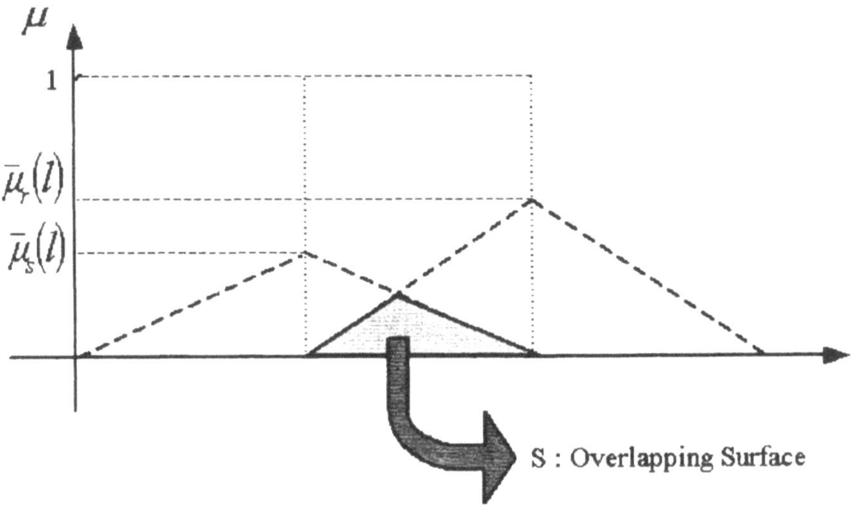

Fig. 8.7 Construction of triangular membership functions and determination of the overlapped surface used to determine the mass function of a double hypothesis

Thus, S depends on $(\mu_r(l)+\mu_s(l))$ and $(\mu_r(l)-\mu_s(l))$ to correctly model the existing ambiguity between r and s. Another condition that S must satisfy is that its value has to be confined to the interval [0,1]. To do that, S is normalised with the maximum value S_{max} of S. From the membership functions of hypotheses H_r and H_s, the mass value assigned to the composite hypothesis $H_r \cup H_s$ is then given by:

$$\tilde{m}(H_s \cup H_r) = \Im(\mu_r(l), \mu_s(l)) = 0.5 \cdot \frac{S(\mu_r(l), \mu_s(l))}{S_{max}} \tag{8.19}$$

From equation (8.19), it is seen that the value of $\Im$ is obtained through the ratio of two surfaces of the same type. So, the representation shape of the surface has no influence on the final results. This also justifies the choice of the triangular shape owning to its computation simplicity. On the other hand, in FCM clustering each class is characterised by a centre (grey level in this study) having $\mu = 1$. Therefore, when moving away from this centre, the membership degree to this class decreases, and at the same time the membership degree to the neighbouring class increases until reaching its core which has a value of $\mu = 1$. Hence, an overlapping of 50 % between the two classes is produced, as shown in figure 8.3a. Because of the normalisation in equation (8.19), the common base (figure 8.7) of the triangular surface S is eliminated in the computation. This means that the distance between the vertexes of the two triangles is not involved in the final result. So, this distance can be set to 1 for simplification. The only important values are $\mu_r(l)$ and $\mu_s(l)$.

After calculating the mass functions for the two images, the combination of the mass functions is achieved using the DS combination rule. The decision procedure for classification purpose consists in choosing the maximum belief as the most likely hypothesis H_i. The decision-making is carried out on simple hypotheses which represent the classes in the images. If we accept the composite hypotheses as final results in the decision procedure, the obtained fusion results would be more reliable but precision would have decreased.

Results

Figure 8.8 is an example of image fusion on industrial images, the size of which is 1000 × 300 pixels after registration. The ultrasonic image exhibits high contrast, but the spatial resolution of the image is poor and the objects are not well separated. The image also reveals a textured background, which represents spatial noise (this may not be very visible if the image is viewed with more than 128 grey levels since the human visual system can only distinguish about 60 grey levels). In contrast, the X-ray image presents good spatial resolution. The small objects are well separated, and their shapes are often close to the reality. There is also little granular noise. However, the contrast of the image is not so satisfactory as in the ultrasonic image. Some features are missing due to the fact that ultrasonic imaging offers higher detection sensitivity than X-ray radiography.

The fusion result through the combined use of FCM and DS theory is shown in figure 8.8c. The original radiographic and ultrasonic images have relatively significant differences in terms of the shape and number of defects. In the radiograph only five defects are clearly visible, with a sixth defect barely visible. In the ultrasonic image, many objects are present. The fusion result in figure 8.8c reveals interesting features. Almost all the objects in both X-ray and ultrasonic images are present. The false negatives (i.e. defects exist but are not detected) can be excluded. The shapes of objects in the fused image are not similar in both images. A compromise has been observed. Further post-processing can be performed to identify the detected objects.

8.5.2 Fusion of Radiographs and Ultrasonic Signal

In this case study, radiographic data is under the form of a 2-D image, and ultrasonic data under the form of one-dimensional signal; their fusion is to be performed at high level. At high level, the elements to be fused are not image pixels, but more complete data entities such as objects or local decisions (e.g. object detection, identification, sizing). As in the case of low level data fusion, the aim is always to improve the precision and reliability of final decisions through the combination of complementary and redundant information from different sources.

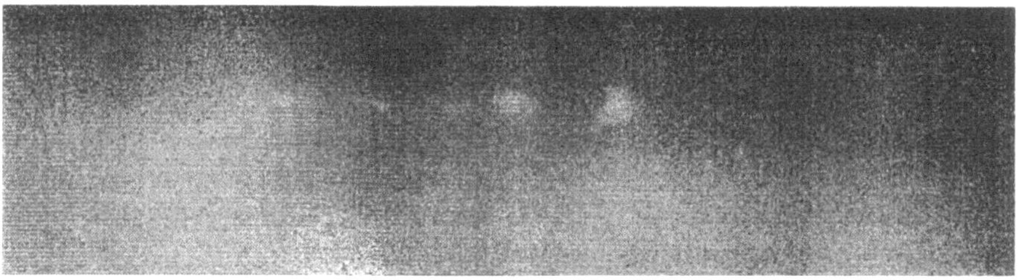

(a)

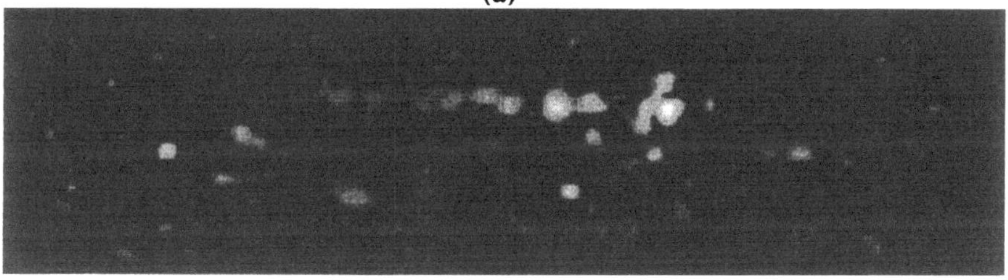

(b)

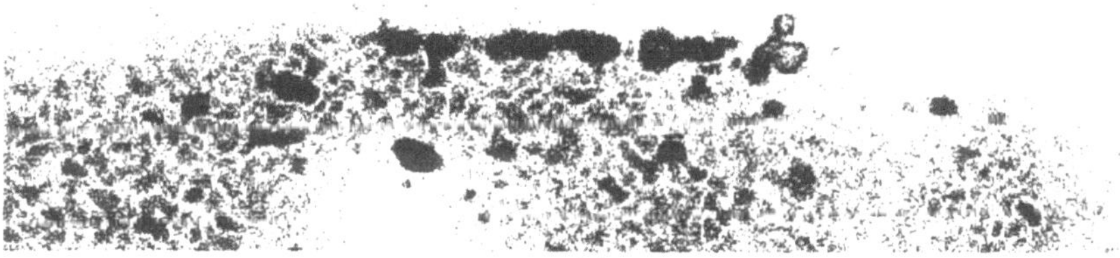

(c)

Fig. 8.8 Illustration of the FCM and DS theory based fusion on a weld piece; (a) original radiograph, (b) original ultrasonic image, (c) fusion result

Figure 8.9 shows a general scheme of high level data fusion. Data from each source (inspection modality) is first handled in a separate manner. Pre-processing aims to reduce artefacts such as noise, and enhance useful edge information. Many available processing techniques can be used for this purpose. The most popular techniques are low-pass filtering, adaptive filtering and restoration. At this stage, registration is also needed as a necessary prerequisite for further processing if non-registration phenomena are involved. The following step often concerns features extraction including features definition and selection, segmentation, or labelling (classification or identification in certain cases). Next, pertinent parameters of the objects are obtained. According to these features, information modelling (e.g. likelihood function in Bayesian approach, mass function in DS theory) is achieved for each object. Combination operators are then used for object fusion. Decision related to problems of detection, identification or sizing is finally made.

In the case of a radiographic and ultrasonic data fusion, processing results for each modality will provide detected objects that can be either true or false defects. When using only one inspection modality, the decision concerning the presence of a defect might not be reliable. Nevertheless, a mass value representing the confidence one has in the detection of each inspection can be defined. The combination rule of the DS theory allows the two confidences to be fused, resulting in higher reliability of the final decision.

In comparison with low-level data fusion, high level fusion is not directly concerned with uncertainty and imprecision of raw data because of processing performed. This could constitute a disadvantage for such fusion methods if the uncertainty and imprecision are better modelled at the pixel level. The major advantage of the high level fusion is that raw data can be pre-processed by experts in a flexible manner and in a more efficient way. As in low level fusion, the difficulty is to define mass functions and automatically attribute masses to objects detected with both inspection modalities. To overcome this difficulty, the pertinent object parameters that should allow the true and false defects to be distinguished should be determined. Objects are fused by combining their masses derived from those parameters. To achieve this, a statistical study on the relation between defect types and feature parameters is to be performed.

X-ray Data Analysis: Data Processing

Statistical study on the relation between defect types and feature parameters requires the use of a large number of different types of defects. As many as 386 defects from digitised radiographic films were used for this purpose. They cover almost all types of weld defects (i.e. porosity, inclusion, lack of fusion, etc.).

A segmentation algorithm based on mathematical morphology was applied on radiographic images to detect as many true defects as possible. A pre-processing is applied to the radiographic image using a median filtering to eliminate background noise. As the image background is not constant, one simple threshold is not sufficient to segment the image. So, an adaptive thresholding method based on the morphological top hat transform was selected [26]. When considering the grey-level image as a geographical relief, where the grey levels are heights, the defects become peaks. The top hat transform consists in selecting these peaks whose height is greater than a given threshold, and whose size is less than another threshold. In this way, defects are well detected but noise may also be detected as useful information, as the latter appears as high and fine peaks. To reduce noise in the segmented image (reference image), a median filter eliminates small and isolated particles, while keeping the nuclei of the significant defects (markers image). Then, the defects are retrieved by dilating the nuclei conditionally with respect to the reference image.

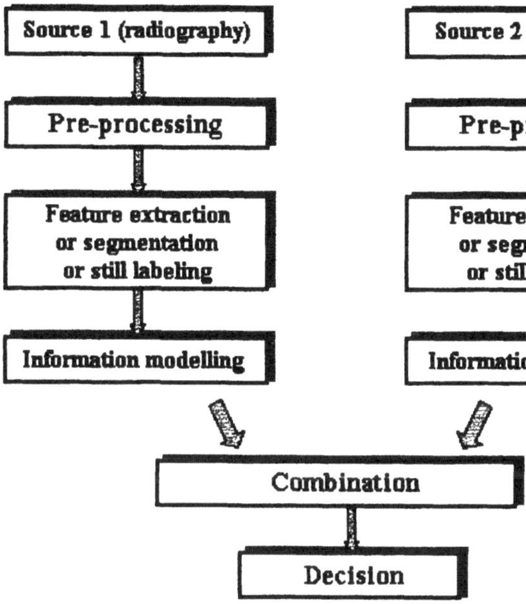

Fig. 8.9 General scheme of high level data fusion

Whatever the segmentation method used, all segmented objects may not be defects, and all defects may not be present in the segmented image. In other words, detection of defects via segmentation is not ideal in the sense that defects may remain undetected and that false alarms may occur. There exist other segmentation methods, but the choice of the segmentation method is often a compromise between reliability (i.e. detection of the higher number of real defects as possible) and percentage between real defects and false alarms. It would always be a compromise between false positives and false negatives. Because image processing is designed to keep in the binary (segmented) image a high number of real defects, many false defects are present. For example, a low value of threshold allows the detection of almost all real defects but at the same time of a lot of false defects, whereas a high thresholding value reduces and even annuls false detection signals but degrades the reliability of the inspection. The reduction and even elimination of false defects is expected by using data fusion. In the present section, image processing leads to a rate of detection greater than >75%. Figure

8.10 shows segmentation result on a radiographic image presenting a lack of fusion.

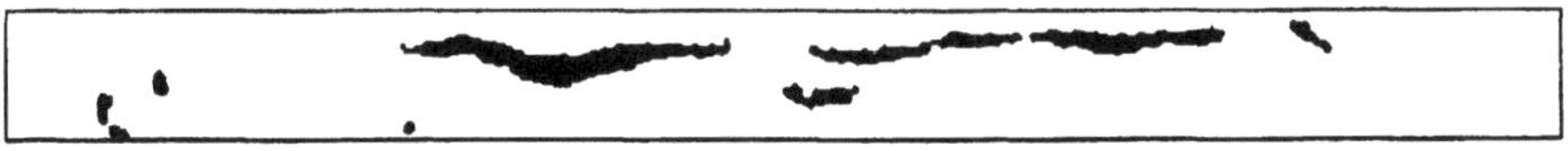

Fig. 8.10 Segmentation of a weld image presenting lack of fusion

X-ray Data Analysis: Definition of Feature Parameters

After segmentation, the detected objects are analysed using feature parameters. In practice, to detect and identify defects in welded pieces human experts essentially use the geometrical shape of the defect (e.g. elongated or rounded), its contrast (i.e. the signal variation of the defect to the background), its position and size. They usually take into account the welding method to estimate the probability of appearance of one type of defect in a specific weld. Here, the segmented objects are characterised by their contrast, area and shape. The area corresponds to the number of pixels included in the object. The contrast is computed from difference between the mean grey-levels of the object and the background. The spatial neighbourhood used for evaluating the mean grey-level of the background is a region of shape annulus. It is obtained by subtracting the segmented object to its dilated (morphological operation) version. The shape is obtained by computing the ratio of Feret diameters. These ones represent the distance between the object shape and a circle, for which the shape factor is equal to 1.

Figure 8.11 shows a statistical study of two defect features, area and contrast. All objects in the segmented image are classified into true or false defects. In terms of DS theory, the frame of discernment is constituted of two hypotheses; H_1 there is a defect, and H_2 there is no defect. The uncertainty between these two hypotheses is expressed as the union of H_1 and H_2. It also represents total ignorance. In figure 8.11, nine regions are defined. Each region represents a percentage of true and false defects as shown in table 8.3. In regions 1 and 2, there are as many true defects as false defects, reflecting the total ignorance on whether objects in these regions are true or false defects. The hypothesis $(H_1 \cup H_2)$ is more suitable to describe this situation. In contrast, the objects in regions R_5 to R_9 are true defects. This means that an object in one of these regions is surely a true defect. So, there is strong belief towards the hypothesis H_1. Region R_4 includes an important proportion of true defects, indicating higher confidence on the hypothesis defect. At the same time, the presence of false defects yields hesitation between these two possibilities. This gives a strong belief that objects in this region are true defects. Finally, 66% of false defects are observed in the region R_3. In principle, this means that there are almost no defects in this region. However, because of the small number of objects observed in this region, the results of the statistical analysis should be less reliable. For this reason, there still exist uncertainties between true and false defects. The above analysis on the choice of hypotheses H_1 and H_2 can be expressed by gradually representing uncertainty between H_1 and H_2:

- P_1 : total ignorance,
- P_2 : strong hesitation but small preference for one of the two hypotheses,
- P_3 : high confidence for one of the two hypotheses with slight doubt,
- P_4 : no uncertainty, it is one of the two hypotheses.

For the propositions P_2, P_3 and P_4, we use P_i(defect) when the proposition favours the hypothesis that a defect is present, and P_i(no defect) when the hypothesis that no defect is present is preferred. Each region in figure 8.11 receives a proposition, as shown in the last column of table 8.3. From these propositions, masses in DS theory are derived, as shown in table 8.4.

In practical applications, borders between each region of figure 8.11 are not so abrupt, since there is no reason that substantially different confidences are committed to two close objects (in terms of area and contrast). To solve this problem, fuzzy logic is a useful tool that ensures smooth transitions between each region. For the present application, three fuzzy subsets are defined for the contrast C (low, medium, high) and for the area (small, medium, large). The associated membership functions are represented by $\mu_{C(j)}$ and $\mu_{S(k)}$ (j, k=1, 2, 3), respectively.

By designating the region associated with the classes C(j) and S(k) by $R_{j,k}$ (figure 8.12), the region R_2 in figure 8.11 corresponds, for example, to $R_{1,2}$ in figure 8.12. Mass functions associated with each region are denoted by $m_{R_{j,k}}$. For an object located in the region $R_{j,k}$ the mass function of the related hypothesis is calculated using:

$$m_{RX}\ (H_i) = \sum_{j,k} \mu_{C(j)}(x) * \mu_{S(k)}(y) * m_{R\,j,k}(H_i) \tag{8.20}$$

The weighting using membership functions in equation (8.20) allows the elimination of abrupt transition of mass functions from one region to another.

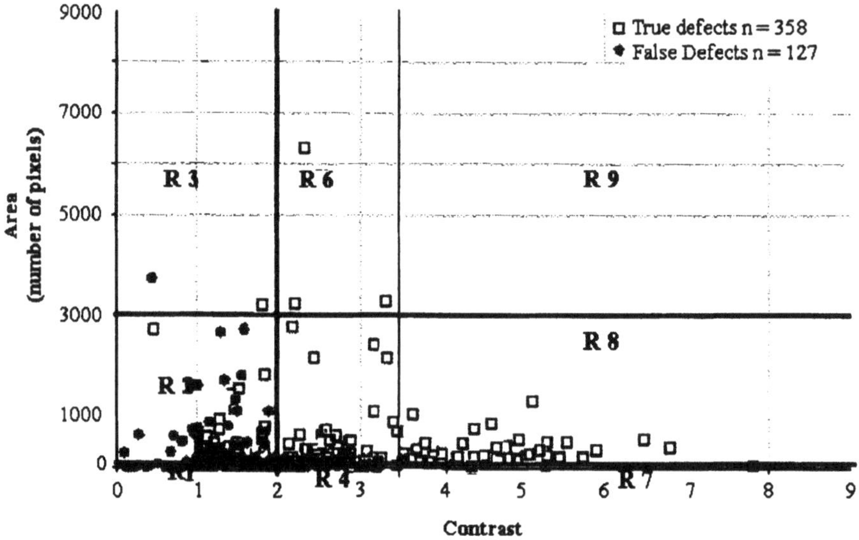

Fig. 8.11 Illustration of the partition of features plane into regions

Table 8.3 Repartition in percentage of true and false defects for different regions

Region	Number of objects	True defects	False defects	Proposition
R_1 to R_2	255	55%	45%	$P_1(H_1 \cup H_2)$
R_3	3	33%	67%	$P_2(H_2)$
R_4	73	83%	17%	$P_3(H_1)$
R_5 to R_9	154	100%	0%	P_4 (H_1)

Table 8.4 Derivation of mass functions from propositions

Proposition	$m(H_1)$ (defect)	$m(H_2)$ (no defect)	$m(H_1 \cup H_2)$ (ignorance)
$P_2(H_2)$	0	0.33	0.67
$P_3(H_2)$	0	0.67	0.33
$P_4(H_2)$	0	1	0
P_1	0	0	1
$P_2(H_1)$	0.33	0	0.67
$P_3(H_1)$	0.67	0	0.33
$P_4(H_1)$	1	0	0

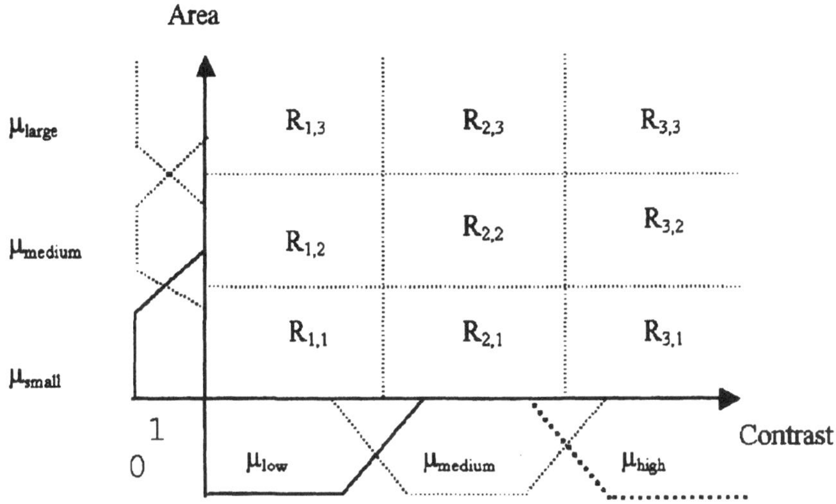

Fig. 8.12 Weighting of features plane using membership functions in fuzzy logic

Ultrasonic Data Analysis

In practical ultrasonic inspections, it is commonly agreed that a high amplitude echo is representative of the presence of a defect. Thus the operator relies solely on the amplitude information of the ultrasonic signal. In the case of weld inspection, signals with an echo amplitude higher than that of a reference hole of a given diameter is associated with the presence of a defect with a high degree of confidence. The amplitude reference (or amplitude threshold) is often taken as twice the background noise, since the propagation of ultrasonic wave in carbon steel is usually not very noisy.

Figure 8.13 shows the division into three intervals of ultrasonic signal amplitudes. The partition was obtained using an amplitude threshold identified after analysing a great number of ultrasonic inspection signals. The interval I_1 represents the low echo amplitudes, and the interval I_3 the high echo amplitudes. The interval I_2 designates the medium echo amplitudes. The interval I_1 (or I_3) represents the situation where certainty of having no defects (or defects) is high. In contrast, higher detection uncertainty exists for ultrasonic signals, the amplitudes of which fall into the interval I_2. In particular, the analysis of ultrasonic echo signals revealed the existence of a specific amplitude value for which a correct detection of defects is highly uncertain.

As for the case of X-ray data, propositions can be associated with each of the three intervals in order to determine mass functions. The proposition P_1 (total ignorance) is associated with the interval I_2, and the proposition P_4 (for hypothesis H_1 or H_2) to intervals I_1 and I_3. Fuzzy logic is used to smooth the abrupt transition between the intervals, . The weighting (smoothed) mass function of a hypothesis is given by:

$$m_{US}(H_i) = \sum_j \mu_{A(j)}(x) * m_{Ij}(H_i) \tag{8.21}$$

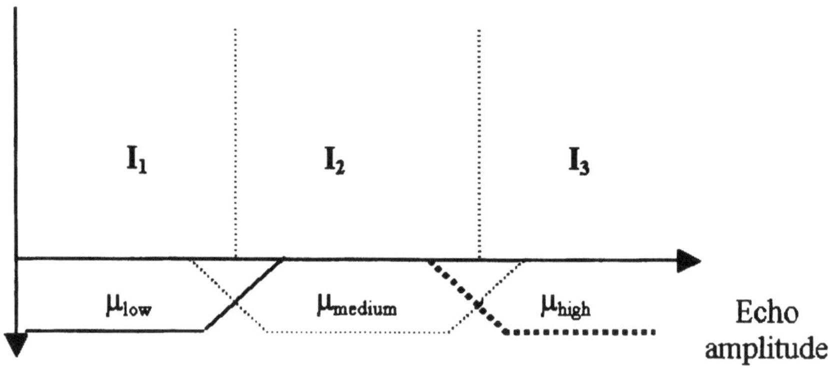

Fig. 8.13 Partition into intervals of ultrasonic signal amplitudes and weighting using fuzzy logic

Fusion of X-ray and Ultrasonic Data

Once the mass m_{rx} and m_{ut} are determined, they are combined using the combination rule in equation (8.3). Decision is made on the combined masses using the maximum belief. Figure 8.14 compares masses related to radiographic inspection with those from X-ray/US fusion. It is seen that when only radiographic data is used, one cannot determine if there is no defect, the mass corresponding to ignorance is large. There is therefore high uncertainty on saying that there is no defect. Concerning the detection of true defects, three situations can be distinguished. Radiographic inspection allows the detection of most volumetric defects. The related masses are fairly great. Moreover, they are close to the masses obtained after fusion with ultrasonic data.

The two other situations correspond to cases of planar defects in which two types of planar defects can be distinguished. The first type is planar defects, which have been divided into small segments during image processing. The second type corresponds to planar defects, the morphological aspect (surface in particular) of which has not been destroyed by image processing. This distinction is important since some planar defects of type 1 have fairly high detection uncertainty. It was observed that masses for the detection of both planar defects are substantially higher after radiographic and ultrasonic data fusion. The ignorance is also largely reduced. The masses increased from 0.5 to more than 0.8 for the first type of planar defects, and from 0.6 to almost 1.0 for the second type of planar defects.

Planar defects are often split into small objects by radiography and image processing operations. Figure 8.10 illustrates such a situation. The split objects have been regrouped into one object after fusing X-ray and ultrasonic data as shown in figure 8.15.

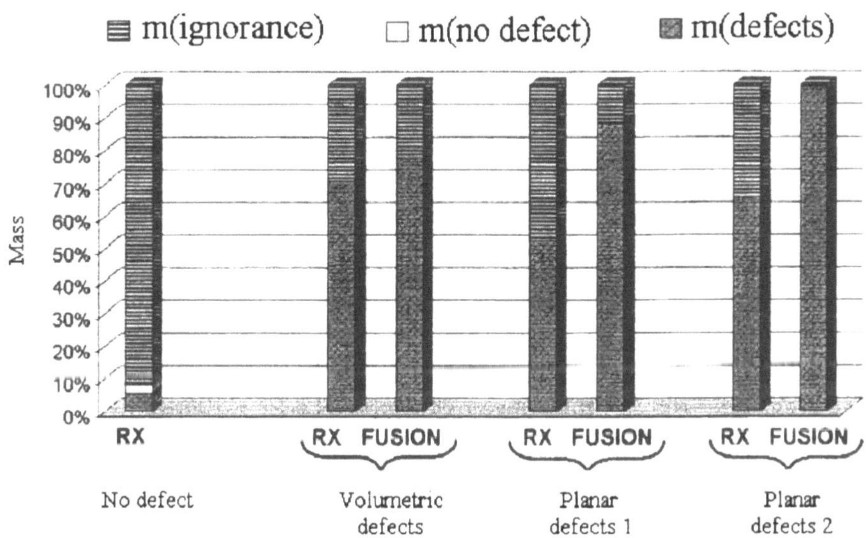

Fig. 8.14 Comparison of masses of X-ray radiographic examination with those of X-ray/UT fusion results

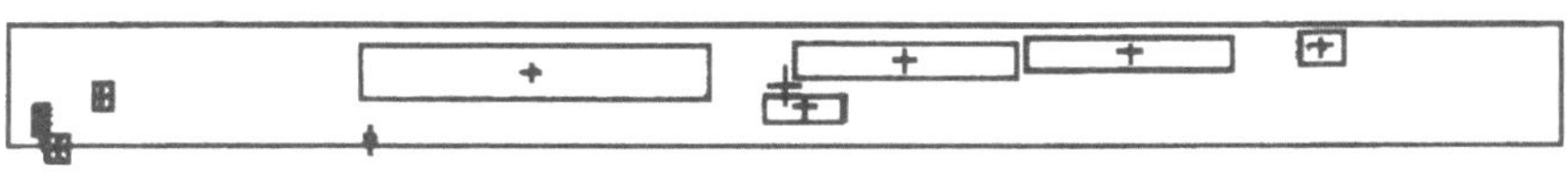

Fig. 8.15 Regrouping split objects by fusing ultrasonic data and the radiograph from figure 8.10

8.6 CONCLUDING REMARKS

Inspection speed, quality, reliability and cost are the key characteristics desired for many NDT problems including weld inspection. The concept of data fusion provides an interesting approach for designing inspection systems meeting these requirements. The idea is to use software solutions that exploit in a synergistic manner, complementary and redundant information from different sources. It was described in this chapter how to substantially improve the dynamic range of X-ray imaging systems, enhancing defect detection sensitivity, without modifying the imaging system. This concept remains interesting even if the X-ray imaging system has already a good dynamic range. Data fusion performs a so-called adaptive imaging by choosing the X-ray energy based on the thickness of the objects to be tested. Further improvement can be expected if image restoration techniques are used to improve the spatial resolution of the X-ray imaging system.

The possibility of improving the reliability of detection of weld defects was also demonstrated by fusing radiographic and ultrasonic data. Although the multimodal NDT is already a reality in the industrial field, it has been, until now, performed in a manual manner. The automation of this manual or mental fusion is fundamental for obtaining objective and reproducible multimodal inspection. The combined use of DS theory and fuzzy logic can make this automation possible. The results of data fusion on real defects illustrated in this chapter have demonstrated the great potential of fusing radiographic and ultrasonic data. Future work will involve validation of data fusion methods using a representative X-ray and ultrasonic database, and comparing fusion results with manual experimental results provided by qualified human operators.

REFERENCES

1. Lundervold A., Storvik G., *Segmentation of brain parenchyma and cerebrospinal fluid in multispectral magnetic resonance images*, IEEE Trans. on Medical Imaging, 1995, 14(2):339-349.
2. Dromigny-Badin A., Zhu Y.M., Grimaud J., Pachai C., Boudraa A., Hermier M., Gimenez G., Froment J.C., Confavreux C., *Use of data fusion techniques for the automated segmentation of multiple sclerosis lesions in magnetic resonance images*, Proc. 8th meeting of the European Neurological Society, 1998, Nice, France, 446.
3. Moy J.P., *Recent development in X-ray imaging detectors*, Nuclear Instruments and Methods, 2000, 442(1-3):26-37.
4. Jacquemod G., Odet C., Goutte R., *Image resolution enhancement using subpixel camera displacement*, Signal Processing, 1992, 26(1):139-146.
5. Odet C., Jacquemod G., Peyrin F., Goutte R., *Improved resolution of 3D X-ray computed tomographic images*, Proc. SPIE Visual Communication and Image Processing Conf., 1990, Switzerland, 1360:658-664.
6. Zhu Y.M., Babot D., Peix G., *A quantitative comparison between linear x-ray sensitive array and image-intensifier systems*, NDT International, 1990, 23(4):214-220.
7. Tanner R., Loh N.K., *A taxonomy of multi-sensor fusion*, J. of Manufacturing Systems, 1992, 11(5):314-325.

8. Ray Smith. C., *Bayesian approach to multisensor fusion*, Proc. SPIE Signal processing, sensor fusion and target recognition, 1992, Orlando, USA, 1699:285-299.
9. Richardson J.M., Marsch K.A., *Fusion of multisensor data*, Inter. J. of Robotics Research, 1988, 7(6):78-96.
10. Denoeux T., *A k-nearest neighbor classification rule based on Dempster-Shafer theory*, IEEE Trans. on Systems, Man and Cybernetics, 1995, 25(5):804-813.
11. Bloch I., *Some aspects of Dempster-Shafer evidence theory for classification of multi-modality medical imaging taking partial volume effect into account*, Pattern Recognition Letters, 1996, 17(8):905-919.
12. Dromigny-Badin A., *Image fusion using evidence theory: applications to medical and industrial images*, PhD thesis, INSA Lyon, France, 1998.
13. Gros X.E., Strachan P., Lowden D., *Theory and implementation of NDT data fusion*, Research in Nondestructive Evaluation, 1995, 6(4):227-236.
14. Dromigny A., Rossato S., Zhu Y.M., *Fusion de données radioscopiques et ultrasonores via la théorie de l'évidence*, Traitement du Signal, 1997, 14(5):499-510.
15. Dromigny A., Zhu Y.M., *Improving the dynamic range of real-time x-ray imaging systems via Bayesian fusion*, J. of Nondestructive Evaluation, 1997, 16(3):147-160.
16. Gros X.E., *NDT Data fusion*, Butterworth-Heinemann, 1997.
17. Abidi M.A., Gonzalez C., *Data fusion in robotics and machine intelligence*, Academic Press, London, 1992.
18. Shafer G., *A mathematical theory of evidence*, Princeton University Press, 1976.
19. Luo R.C., Kay M.G., *Multisensor Integration and Fusion in Intelligent Systems*, IEEE Trans. on Systems, Man and Cyber., 1989, 19(5):901-931.
20. Bogler P.L., *Shafer-Dempster reasoning with applications to multisensor target identification systems*, IEEE Trans. Systems Man Cyber., 1987, 17(6):968-977.
21. Waltz E.L., Buede D.M., *Data fusion and decision support for command and control*, IEEE Trans. on Systems, Man and Cyber., 1986, 16(6):865-879.
22. Zadeh L.A., *Fuzzy sets*, Information and control, 1965, 8(3):338-353.
23. Li H., Yang H.S, *Fast and reliable image enhancement using fuzzy relaxation technique*, IEEE Trans. on Systems, Man and Cyber., 1989, 19(5):1276-1281.
24. W. Pedrycz, *Fuzzy sets in pattern recognition: methodology and methods*, Pattern Recognition, 1990, 23:121-146.
25. Bezdek J.C., *A convergence theorem for fuzzy isodata clustering algorithm*, IEEE Trans. Pattern Analysis and Machine Intelligence, 1980, 2:130-142.
26. Serra J., *Image analysis and mathematical morphology*, Academic Press, New-York, 1984.

This study was supported in part by the european programme Brite-Euram III. Sincere thanks are due to A. Dromigny-Badin, O. Dupuis, L. Bentabet and A. Boudraa for their contributions in the implementation of the fusion techniques.

9 NDT DATA FUSION IN CIVIL ENGINEERING

Uwe FIEDLER
Fraunhofer Institute,
Dresden, Germany

9.1 INTRODUCTION

Support construction elements in civil engineering are usually made of reinforced concrete. Cracks in pre-stressed steel bar bundles, which can be induced by hydrogen, may critically diminish the safe load of bridges. Non-destructive testing methods are required that are capable of detecting cracks as early as possible. These defects progressively reduce the effective diameter of the steel bar bundle until complete breakage. Dependent on special requirements, verification by drilling or observation of crack growth is to be decided upon from non-destructive measurements.

Essentially for interpretation, signals of defects in reinforcement bundles have to be separated from various artefacts. Artefacts can be caused by crossbars, overlaying reinforcements, or ceasing bars. Measuring conditions vary depending on the applications. Indeed, different material parameters (e.g. concrete, steel), and configurations (e.g. number, diameter, ordering of bars, magnetic characteristics of measuring system) are used in civil engineering, which make structural integrity assessment difficult. In addition, because tests have to be specifically prepared, full test automation is inadequate.

Siempelkamp GmbH develops a measuring device for detection and characterisation of defects to evaluate the safe load of bridges. They are industrial end-users of the measuring technique described in this chapter, and provided continuous practical experimentation and feedback for the progress of the project.

9.2 THE MEASURING APPROACH

9.2.1 Measuring Principle

The magnetic measuring approach used is based on the flux leakage principle [1-3]. To enhance spatial resolution, a matrix of 8 × 8 independent sensors has been designed (figure 9.1), which was prototypically realised as eight sensors in a single row. Distributed sensors are essential to separate signal components arising from spatially distributed elements as crossbars, which are located above the measuring objects. By moving the magnetic row in parallel along the *x*-axis of the measuring object (longitudinally), matrix like measurement could be simulated.

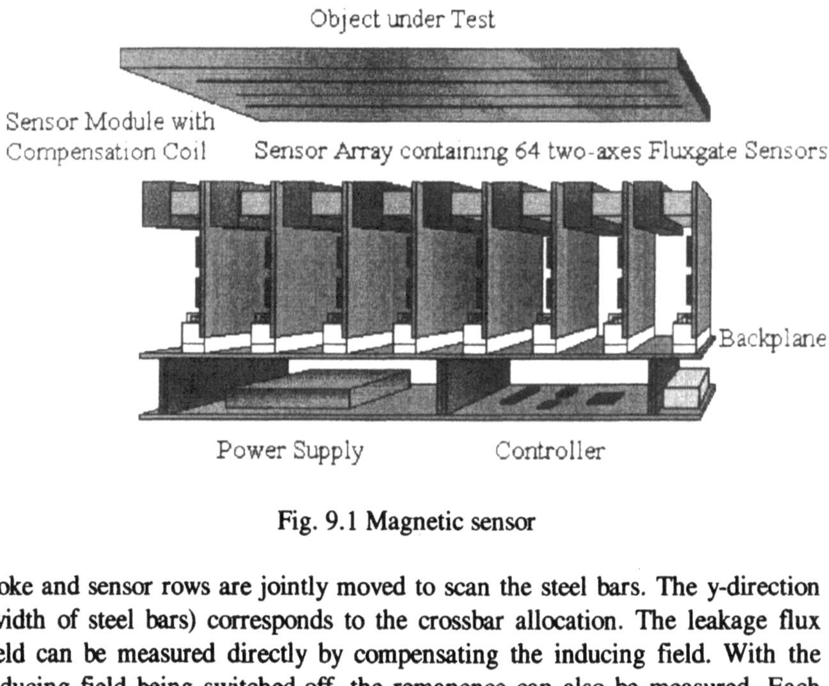

Fig. 9.1 Magnetic sensor

Yoke and sensor rows are jointly moved to scan the steel bars. The y-direction (width of steel bars) corresponds to the crossbar allocation. The leakage flux field can be measured directly by compensating the inducing field. With the inducing field being switched-off, the remanence can also be measured. Each sensor is capable of measuring magnetic fields in two directions (H_x, H_z), both for leakage flux and remanency mode. Digital signal processing assigns time, location and sensor signals during scanning to compensate for limited scanning and sensing speed.

9.2.2 Measuring Parameters

The measuring approach delivers four types of information. Switching the inducing field on or off, either leakage flux or remanency can be measured. Both techniques deliver information in two measuring directions (H_x, H_z). The resulting signals are measured as voltages, and are partly complementary, dependent and even contradictory, therefore requiring information fusion. To develop the approach, significant measuring parameters were analysed:

- magnetising current: I (in ampere),
- height of sensor above reinforcing steel bar: z (in cm),
- size of crack: r (in mm).

In practical use, the height of the sensors and crack size are unknown parameters that have to be inversely computed from measured voltage signals. For I = 2A, z = 10cm, r = 5mm, the defect indication in the various signals resulting from eight parallel measuring sensor columns can be plotted (figures 9.2-9.5).

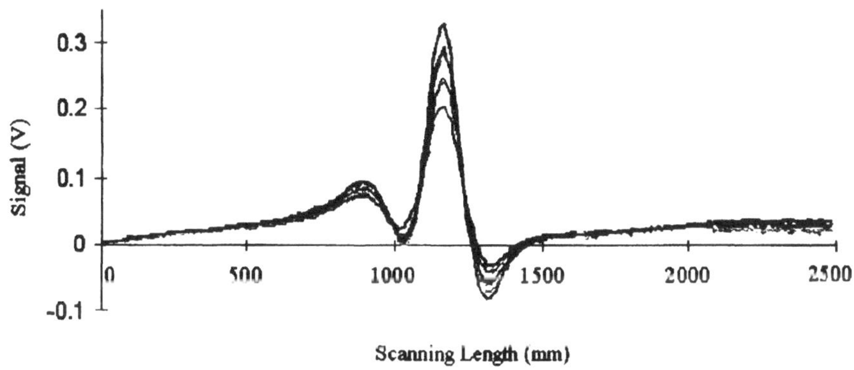

Fig. 9.2 Leakage flux *vs.* scanning length along the *x*-direction; a maximum peak with two symmetric minimum peaks indicate a crack

When measuring in the *z*-direction, the signal shape changes significantly, but also indicates the presence of a crack with great sensitivity. This is a source of complementary information. Remanency measurements in both directions should resemble the corresponding structures obtained with leakage flux. However, practical measurements show variations. This gives rise to the possibility of exploiting these complementary measurements, thus limiting defect detection from multiple decision indications.

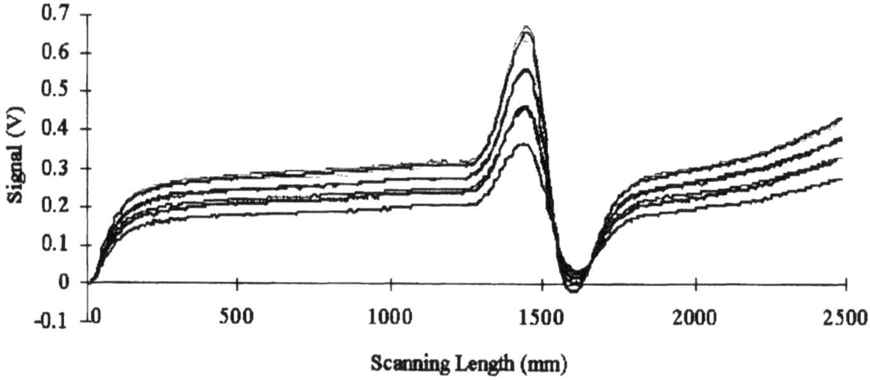

Fig. 9.3 Leakage flux *vs*. scanning length along the *z*-direction; the maximum peak indicates a crack

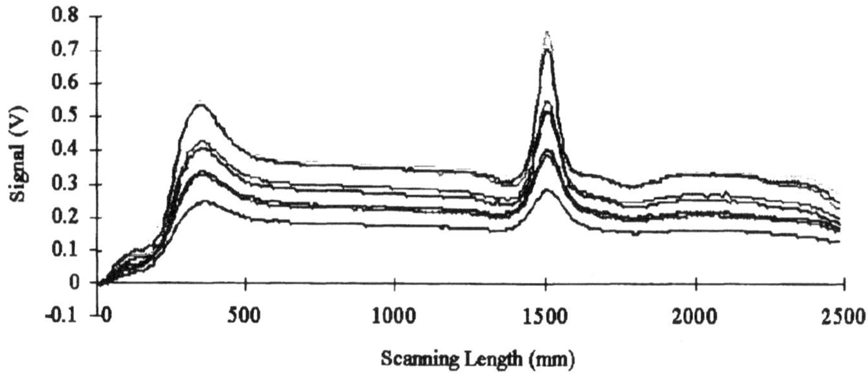

Fig. 9.4 Remanency *vs*. scanning length along the *x*-direction

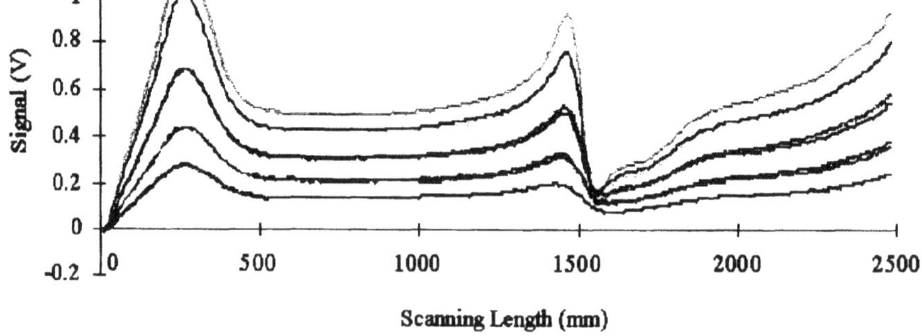

Fig. 9.5 Remanency *vs*. scanning length along the z-direction

9.2.3 Modelling and Data Pre-processing

The probe design as a sensor matrix causes various inherent disturbances. The distribution of sensor calibrations as well as dynamic effects (e.g. measurements of sensor/reinforcement arrangement) can be taken into account to reduce stochastic fluctuations. Physical modelling of relationships between magnetic measurement and material parameters has to be considered. Investigations were meant to deal with several aspects in measurement and testing such as signal separation, information contained within the signal, testing parameters to be adjusted, signal pre-processing operations, and expected results.

The separation between part of the signal containing defect indication and part of the signal related to crossbar is of major importance. Since crossbars are located in irregular distances, the approach to be designed should be able to integrate secondary effects if required (e.g. lengthwise overlapping of steel bars, slack reinforcing bars, etc.). Both leakage flux and remanency signals, both in x and z-direction, contain information useful for defect detection. Exact testing rules are required to define variation of the magnetising current I and the height of sensor above reinforcing steel bar z. Matrix based data fusion should compensate for various stochastic measurement effects. Systematic effects caused by crossbars should be separated at the signal level using magnetic model computations. Analysis of single signals (i.e. signal sequences for varying parameters) involves correlation of these signals with calibration signals (e.g. signature of signals from typical defects). Thus, allowing qualitative evaluation of signal shapes. Each correlation characterises an incomplete, non-sufficient feature supporting or refuting the presence of a defect. As a result, the final decision is derived from the fusion of all these partial decisions for the measuring approaches applied.

9.3 DECISION SUPPORT

A user can decide for the assistance level depending on his needs. Beginning with the basic level of acquiring, visualising and storing data, one can request decision support on a progressing scale of complexity. The implementation of this design can proceed with practical needs. For measurement, the conventional interaction is preserved as the elementary level of decision support. Screening is based on sensitive observation of signals with questionable content. Without claiming for sufficiency, single approaches are applied that must not be specific but especially sensitive. Artificial effects are not yet separated, but the data stream can be compressed before storing. Interactive defect detection is achieved through a hierarchical presentation of an ensemble of measuring curves. Detailed information (i.e. original data) is available on demand. Pre-processing is performed up to the required level, splitting and reconfiguring signals under different views to identify essential parts in the signal. For single approaches, qualitative evaluations of the shape of the signal are integrated to an overall evaluation, which contributes in significance and creditability towards the single evaluation contribution. This results in a quantitative indicator, while the operator remains responsible for the final decision. Automatic evaluation of measurement series without interaction allows automatic defect detection. Finally, to facilitate defect characterisation, defect parameters such as the remaining bar diameter are quantitatively characterised and monitored over a time period by a regression approach using a neural network.

9.4 MODELS OF INFORMATION FUSION

From the application viewpoint, the detection of defects as well as the separation of artefacts from magnetic measuring data can be supported with various methods. Unfortunately, none of them is sufficiently decisive. This is due to important variations of material configurations and measuring parameters. Thus, information fusion based on multiple sensors and multiple approaches is necessary. It can be implemented in dependence on specific application requirements. The quality of decision of based on multi-signal approaches is expected to outperform decision made considering a single method. The analysis of multiple signals offers improved accuracy. From the combination of several signals to a comprehensive approach, the necessity to integrate data from different sources arises. The various approaches can be seen partly complementary, redundant or even contradictory. Depending on the requirements, fusion of such heterogeneous information is performed either at the data, feature or (partial-) decision level.

At data level, fusion of redundant information from each sensor compensates for various effects. This includes sensor variations, movement correction, spatial recognition of the measuring object and separation of artefacts, whose influence can be eliminated from the original signal. Furthermore, resulting signal sets from the variations of measuring parameters (e.g. magnetising currents) can be compressed once more. This approach results in aggregated signals which can be seen as features of the measuring object.

The various aggregated signals are correlated with calibration signals, which can stand for prototypical defects or normal ranges, at the feature level. Furthermore, the measuring window can be split into - eventually overlaying - sub-ranges, which are to be re-aggregated after each individual evaluation to take into account different information. This approach would count for discrete scanning ranges, which could partly reduce defect indications. It delivers correlation coefficients as symptoms that either support or refute the presence of a defect.

At decision level, the contributing symptoms are to be considered as partial decisions and therefore are fused, taking into account different degrees of significance or creditability. It provides the final decision. For each practical application, defects are detected following a sequential order of decisions as described next.

- *Measurement planning*
  Suitable measurements (e.g. leakage flux, remanency) are selected and the variation of several parameters controlled (i.e. magnetising current, height of sensor above reinforcing steel bar). A minimum height of sensors above reinforcements, and respect of crossbar influences, should be defined for optimum signal-to-noise ratio.
- *Redundant measurements with magnetic matrix sensor*
  With measurements in a multi-dimensional signal vector, pre-processing of sensor noise refers to individual sensor characteristics.
- *Data fusion*
  The data pool is compressed; matrix based pre-processing refers to relations between single sensors to compute aggregated signals that weight individual information content.
- *Screening*
  Having selected a sufficiently sensitive signal, thresholding delivers indications of possible defects. This enables further compressing of the original data by disregarding irrelevant signals.
- *Signal evaluation*
  Hierarchical presentation of signals to the operator contributes to additional data compression. Signal details are examined until valid decisions are derived. Evaluation refers to signal shapes; more specifically, to a correlation of signal shapes with actual and calibration signals.
- *Decision fusion*
  The final decision is composed of individual contributions from single signal evaluations.

9.5 DECISION FUSION

9.5.1 Motivation

Generally, decision fusion refers to the integration of heterogeneous data from distributed sources. Knowledge based approaches are proved the most promising [4-7]. They are capable of formalising various *a priori* know-hows about sensor data. For example, correlation of signal indications with material properties. Decision fusion requires distributed and hierarchical knowledge management. It should exceeds the capabilities of purely mathematical or neural classifiers by handling uncertainty both in data measurement and data interpretation.

9.5.2. Data Analysis

Various approaches have been investigated regarding their potential contributions to defect detection and artefact separation. An empirical strategy was selected to vary measuring parameters and observe different resulting signal characteristics. Inversely, conclusions were drawn on how to trace back such monitored experimental behaviour to the roots, i.e. the responsible configuration of steel bars and defects. Among the experimental variations described below the following contributions were observed.

- *Separation of cross-bars*
  Crossbars are the cause of the most important artefact. The approach first consists of separating by analytical modelling the effect of crossbars from the original signal. Next, to complementarily investigate this separation on the decision level. Once, the eliminated signals have been identified, additional effects can be separated without taking into account any cross overlaying, and finally validating cross-bar separation. The quality of the separation operation depends whether calibration was performed under the influence of crossbars or not. If not, leakage flux along the x-direction is sufficient to separate crack areas, but only for specific magnetising currents.
- *Variations of crack width and sensor height above reinforcing bars*
  Remanency measurements, especially along the x-axis, are independent of the two parameters previously mentioned. When correlating with prototypical cracks, detection of crack shapes - regardless of different crack extensions - is possible. Leakage flux measurements can also help, but only for selected sensor heights.
- *Variation of crack width and magnetising current*
  Both remanency measurements reflect changes of the magnetising current in the signal shapes. Minor contributions can also be expected from leakage flux along the z-direction.
- *Investigation of special effects as bar ends and bar bundles*
  In practice, defect detection must neglect bar ends, which cause deviating signals. It must however tolerate different and usually unknown configurations of steel bar bundles. These further obstruct systematic detection approaches based on correlation with signals from prototypical defects. All signals can contribute to the separation of bar ends, with minimum significance of the leakage flux along the x-direction.

These investigations helped deciding which signals must be correlated to reach partial decisions, and how the resulting symptoms must be weighted for efficient decision fusion.

9.5.3 Feature Extraction

Features are extracted by correlating various types of information for varied measuring parameters (e.g. remanency and leakage flux along the x and z directions) with signals calibrated on prototypical defects or on defect free areas. The correlation coefficients are computed according to equations 9.1-9.4, and range from -1 (completely anti-correlated) to +1 (completely correlated), with 0 for uncorrelated.

Covariance:

$$\frac{1}{n}\sum_{i=1}^{n} (x_i - \bar{x})(y_i - \bar{y}) \tag{9.1}$$

where x, y is the information and calibration vectors $\bar{x}$, $\bar{y}$ are the corresponding mean values.

Variance 1:

$$\frac{1}{n}\sum_{i=1}^{n}(x_i - \bar{x})^2 \tag{9.2}$$

Variance 2:

$$\frac{1}{n}\sum_{i=1}^{n}(y_i \quad \bar{y})^2 \tag{9.3}$$

Correlation coefficient:

$$\frac{\text{Covariance}}{\sqrt{\text{Variance 1} \times \text{Variance 2}}} \tag{9.4}$$

The resulting correlation coefficients are to be fuzzy modelled to deliver indications supporting or refuting the presence of defects, and combined to reach a final decision.

9.5.4 Fuzzy Modelling

Coping with all the different unwanted effects, final decisions are made by propagating constraints from less specific screening investigations along further validations and refinements. Signal analysis starts with crossbar eliminated signals. Using remanency measurements, a screening operation sensitively searches for any suspicious indications. This search is complemented by leakage-flux measurements. High correlation means high indication of the presence of a defect. Specificity is increased by a first validation, which exploits remanency measurements that refer to varied magnetising current. Again, high correlation means high indication supporting the presence of a defect. Specific effects such as bar ends are separated. In this case, high correlation means low defect indication. The derived results are validated using original signals (e.g. leakage flux along the *x*-direction) with crossbar impacts to count for uncertainties in crossbar separation. The abbreviations listed in table 9.1 are used for input variables. The variables were measured and pre-processed, i.e. noise was eliminated, matrix data was fused and correlated.

Table 9.1 Signal description of the input variables used in the fuzzy approach and indicated in figure 9.6

Input Variable	Signal
Korrelation Rx	Remanency along *x*-direction
Korrelation Rz	Remanency along *z*-direction
Korrelation Sx	Leakage flux along *x*-direction
Korrelation Sz	Leakage flux along *z*-direction
Korr_Rx_Ivar	Remanency along *x*-direction (*I* varied)
Korr_Rz_Ivar	Remanency along *z*-direction (*I* varied)
Korr_Sx_Ivar	Leakage flux along *z*-direction (*I* varied)

Table 9.2 Content description of the intermediate variables used in the fuzzy approach and indicated in figure 9.6

Intermediate Variable	Content
Sensitivity	Sensitive screening for defect detection
Komplementaer	Less sensitive, complementary defect indicator
Screening	Overall intermediate defect screening indicator
SpezScreen	Refined screening indicator with increased specificity
Ivar_Indikator	Combination of *I*-varied signals
Ende-Indikator	Indication for bar end
Final Screen	Screening indicator, bar ends eliminated.

Fuzzy modelled and differently weighted, these input variables are combined in intermediate variables (table 9.2) to be fused for the final result noted VALID1 in figure 9.6 (Validated Screening Indicator). A fuzzy modelling tool *fuzzyTECH* of Inform GmbH is used, which allows the design of hierarchical fuzzy systems (figure 9.6) [8].

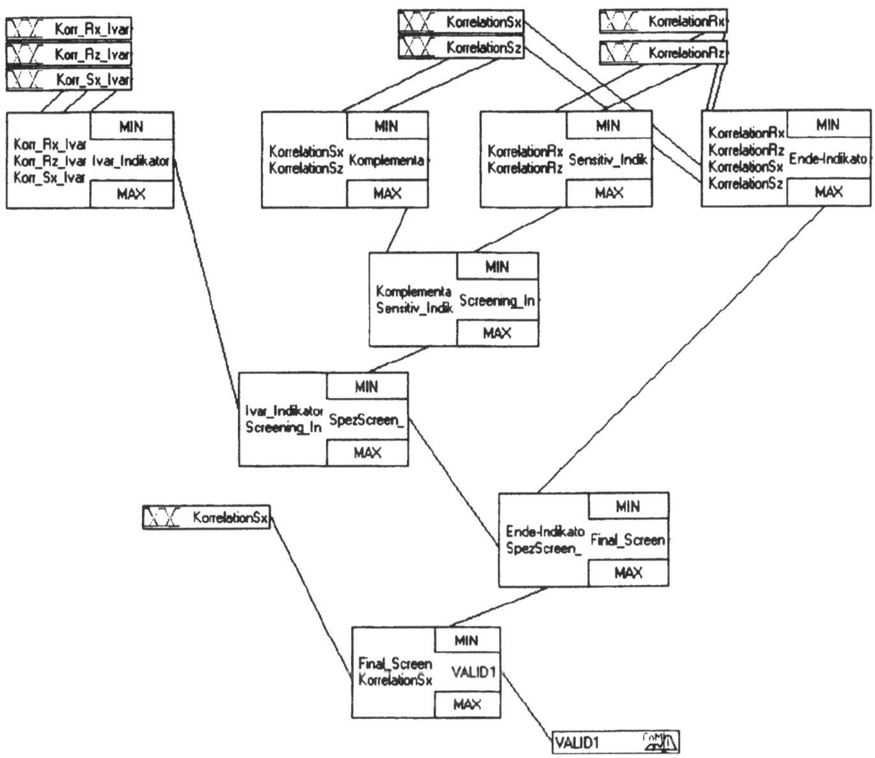

Fig. 9.6 Schematic of the hierarchical fuzzy system used for decision fusion

9.6 CONCLUSION

The developments within the project described in this chapter were part of a feasibility study carried out to outline the potentials of an innovative magnetic measuring technique. Analysis of the requirements led to the fusion of various data sets. Signal analysis, including information fusion, has been designed accordingly. Physical prototyping and application will take place in a follow-up project involving industrial project partners. This project will bridge the gap between basic research and industrial development.

From the implementation viewpoint, the information fusion module has been designed with *fuzzyTECH* for module exchange with *LabVIEW*, a software of National Instruments. Designed as an add-on, the user can easily tailor information fusion to meet specific requirements.

A first prototype of the magnetic sensor has been tested. Software modules were designed and tested off-line with experimental data, and led to promising results. Still, some problems such as information fusion at the data level have to be solved. Thus, further investigations on the information content distributed across the sensor matrix and their dedicated implementation are required to solve them. Nevertheless, the prototype approach outlined the requirements to complete the measuring and signal analysis system. Furthermore, it must be adapted for specific practical applications.

REFERENCES

1. Kusenberger F.N., Barton J.R., Ferguson G.A., *Magnetic Inspection of Reinforcing Steel Rods in Prestressed Concrete*, U.S. Patent N°4573013, 1986.
2. Sawade G., *Quantitative Aspects of the Detection of Ruptures of Prestressed Steels in Concrete Beams using the Magnetic Flux Leakage Measurement Method*, Otto-Graf Journal, 1991, 2:287-310.
3. Sawade G., Straub J., Krause H.J., Bousack H., Neudert G., Ehrlich R., *Signal Analysis Methods for the Magnetic Examination of Prestressed Elements*, Proc. of Int. Symp. Non-Destructive Testing in Civil Engineering, 1995, Berlin, Germany, 2:1077-1084.
4. Bedworth M.D., *Defining decision-level fusion rules using small samples*, Proc. SPIE Conf. on Sensor Fusion: Architectures, Algorithms, and Applications, 1999, Orlando, USA, 3719.
5. Karlsson B., Karlsson N., Wide P., *Reliable Safety Systems with Fuzzy Sensor Fusion in Industrial Robot Applications*, Proc. Int. Conf. on Multisource-Multisensor Information Fusion, 1998, Las Vegas, USA, 353-360.
6. Goebel K., Agogino A.M., *Fuzzy sensor fusion for gas turbine plants*, Proc. SPIE Conf. on Sensor Fusion: Architectures, Algorithms, and Applications, 1999, Orlando, USA, 3719:52-61.
7. Duflos E., Vanheeghe P., Borne P., *Fuzzy fusion operator for mines detection*, Proc. IEEE Int. Conf. on Systems, Man, and Cybernetics, 1998, 2:1156-1161.
8. Inform GmbH Aachen, *FuzzyTECH 5.1*, Reference Manual, 1998.
9. *Magnetischer Matrixsensor*, Project-N°2695/469, Sächsische Aufbaubank, Final Report, 1999.

This chapter refers to a project [9] funded by Sächsische Aufbaubank on behalf of Sächsisches Staatsministerium für Wirtschaft und Arbeit under grant 201871 and performed in collaboration of Siempelkamp GmbH Pirna, MIT Dresden GmbH, Fraunhofer Institutes EADQ and IMS-II Dresden, SMT & Hybrid GmbH Weißig. The author acknowledges discussions with Prof. Lehmann of MIT, for his continuous help concerning magnetic foundations, Dr. Gampe of Siempelkamp GmbH for his praxis-oriented project leadership as well as colleagues at EADQ Dresden, involved in designing and realising the measuring device.

10 NDT DATA FUSION FOR IMPROVED CORROSION DETECTION

David S. FORSYTH, Jerzy P. KOMOROWSKI
NRCC, Institute for Aerospace Research
Ottawa, Canada

10.1 INTRODUCTION

The problem of corrosion in aircraft structure has traditionally been considered a simple economic issue. However, in 1995 Hoeppner *et al.* presented a review of corrosion and fretting related accidents [1]. Their well-documented conclusion that corrosion in aircraft is a safety concern has not swayed the opinion of many who still refuse to accept this position. Independent research in Australia [2] and in Canada [3-4] strongly supports the need to consider corrosion as a threat to safe operation of ageing aircraft and not just as an economic burden. In response to these concerns, and due to rapidly rising costs of fleet maintenance, research is being performed in a number of organisations include corrosion damage in damage tolerance calculations.

Structural models of corrosion and its interactions with fatigue are currently under development, therefore it is impossible to know exactly what resolution and precision of measurements is required from NDT. For example, lap joint specimens pre-corroded to less than 5% thickness loss have demonstrated nearly 50% reduction in fatigue life [5]. This is due to increased local stress concentrations caused by corrosion, which reduces durability and accelerates crack growth. This amount of thickness loss is roughly identical to the sheet thickness tolerance, and provides a serious challenge to conventional NDT technologies. Data fusion is one of the key elements required to meet this challenge successfully.

10.2 EFFECTS OF CORROSION ON AIRCRAFT STRUCTURE

Wallace and Hoeppner proposed detailed definitions of corrosion types that occur in aircraft structures [6]. Three common structural elements subject to hidden corrosion are lap splice joints, fastener holes in thick sections such as wing planks, and internal structural elements. While these categories are not inclusive of all aircraft structure, they are general enough to encompass many of the problems due to corrosion reported on airframes.

10.2.1 Corrosion in Lap Splice Joints

The most common type of corrosion in lap joints is general attack on faying surfaces. This can occur when the adhesive bond or sealant between the layers breaks down, allowing moisture ingress. General attack is often accompanied by pitting corrosion. In severe cases, general attack evolves into exfoliation corrosion. Each type of corrosion has traditionally been characterised in terms of thickness loss of the original material. However, the effects of different modes of corrosion on lap joint fatigue life are not well understood. Brooks *et al.* have shown that the topography of the corroded faying surface can severely reduce the fatigue life of a joint [7].

Another key mode of damage caused by corrosion in lap joints is pillowing. Pillowing is a deformation of the layers of the joint between the fasteners. This deformation is due to the fact that the corrosion products, aluminium oxides, have a much greater volume than the original material. Changes in the stress state of the lap joint due to pillowing modify locations, orientations, and aspect ratios of cracks emanating from the rivets [4]. These are traditionally believed to be the life-limiting damage mode. Pillowing stresses can be large enough to cause cracks that are environmentally assisted cracks under sustained stress [8]. Finally, pillowing stresses can cause failure of the rivets themselves. During disassembly of service-retired aircraft specimens at the Institute for Aerospace Research, many examples of cracked and failed rivets have been found [9]. Often, the failed rivets are still in place; held by corrosion, mechanical interference, or paint. This makes detection of these failed rivets by visual inspection almost impossible.

10.2.2 Corrosion in Thick Sections

Thick aluminium sections such as those used in aircraft wing skins are less vulnerable to the loss of small amounts of material to corrosion. They are susceptible, in many cases, to exfoliation corrosion starting in fastener holes. Fasteners in these structures are often made of a different material, such as cadmium plated steel, and when moisture penetrates through the opening, galvanic corrosion can occur at the point of contact between the dissimilar metals. The intergranular progress of such corrosion along aluminium grain boundaries that run parallel to the surface of the wing skin will lead to delamination of the thin layers of aluminium that is known as exfoliation corrosion. It is desirable to detect corrosion pits at the surface of the hole or intergranular corrosion before they progress to the exfoliation stage. This is not an easy task that usually requires removal of the fastener. Ultrasonic thickness measurements have generally been successful in such situations. Cracking in fastener holes in these structures is also often a concern, and there may be applications for data fusion in visualising simultaneously corrosion and cracks.

10.2.3 Corrosion on Internal Structural Components

Surface corrosion occurs on exposed surfaces of internal structures where cladding and protective coatings have been removed or damaged, or where impervious paint coatings have retained water. In this case, corrosion products often separate the paint from the metal creating surface bumps that may be detectable using visual or enhanced visual methods, provided the parts are accessible and can be cleaned for inspection. The pitting created by this type of corrosion can extend into stress corrosion cracking.

These types of components are usually inspected visually during depot maintenance. Boroscopes can be used where there is limited access. These types of inspections do not provide results that can be easily recorded, registered, and processed as required to implement data fusion. In the future, enhanced visual (EV) inspection methods such as edge-of-light (EOL) may be used for corrosion detection on exposed surfaces, perhaps in combination with eddy currents (EC) to detect cracks growing from corrosion pits. In this scenario, the registration of the EOL and EC data onto a complex surface, and fusion of the data to relate the existence of cracks to corrosion would be of benefit to structural engineers who have to make repair/rework decisions.

10.3 DATA FUSION FOR NDT OF AGEING AIRCRAFT STRUCTURE

The advanced analyses of the structural effects of corrosion require a sophisticated use of NDT techniques. New structural models will require quantitative input in the form of a quantitative measure (direct or indirect) of corrosion and its effects. For example, in the case of lap joint structure, the metrics of interest will likely include pillowing deformation, material loss per layer, corrosion morphology, crack size and location, and rivet condition. It is not likely that one NDT technique alone will provide all the information required for a complete assessment of the structural integrity of the lap joint. Information fusion can bring together the results of different inspection apparatus to provide sufficient information required for structural integrity assessment.

The generic steps required to perform data fusion on NDT data are simple: data is collected, pre-processed, and then registered on a common co-ordinate system. At this stage, data fusion algorithms can be executed. It is important to note that the steps of pre-processing and registration are value-added steps. Even before any data fusion operation has been performed, data sets from disparate sources have been brought together on one software platform, and registered on a common co-ordinate system. This is a significant improvement over most current practices for handling multivariate data, and greatly facilitates the use of databases for maintenance planning. It also allows improved inspector interpretation by making comparison between NDT data much simpler. Often the pre-processing step can be used to transform a single data source from the NDT domain to a quantitative measure; for example, ultrasonic scans can be transformed from time-of-flight to thickness, or EOL images can be transformed from brightness levels to images of maximum pillowing deformation [10].

The final data fusion algorithm used will be specific to the application. Development of such algorithm will only be cost-effective for repetitive inspection situations, such as the lap splice joint, which is a common structural element on a number of ageing aircraft models. However, the preliminary steps of data handling, pre-processing, and registration are likely to become more commonly used as fleet maintenance practices are automated, thus reducing implementation costs of data fusion in practical situations.

The term 'data fusion algorithm' could refer to any of a bewildering array of choices ranging from simple visual overlay to probability-based methods such as Bayesian inference or Dempster-Shafer evidential reasoning [11]. The technology exists to easily implement visual overlay and pixel level mathematical operations for data fusion, which will be demonstrated in the following sections. Whether this may provide sufficient reliability and sensitivity for the damage tolerance/durability assessment models being developed for ageing aircraft is still uncertain. If not, more complex data fusion methodologies may be required.

10.3.1 Data Fusion for NDT of Lap Joints

The outputs required from the application of data fusion to NDT of aircraft fuselage lap joints must be supplied by the structural engineering community. While requirements for non-destructive examination of lap joints are not yet fully developed, it can be reasonably assumed that metrics of interest will include some or all of the following:

- pillowing deformation,
- joint thickness by layer,
- corrosion topography (e.g. pit sizes, distribution),
- crack location and size.

A two-pass inspection has been proposed by different organisations as the most efficient for NDT of ageing airframes [10,12]. This would entail one rapid technique such as D-Sight or EOL followed by a slower, but more sensitive technique such as conventional or pulsed eddy currents where indicated. The key rationale for this approach is to minimise the cost of maintenance while preserving safety. The loss of operating time on the aircraft is usually the largest component of maintenance cost, and even with modern scanning systems eddy current inspection of all the joints and doublers on a large airframe is a huge undertaking. Enhanced visual NDE is generally an order of magnitude faster than scanned EC, and this approach minimises the areas, which need to be covered by slower NDT techniques.

During the first inspection pass, if historical inspection data exists, a comparison could be performed online to see if the pillowing deformation in the joint had changed. Any change indicating possible corrosion would trigger a secondary inspection with eddy current. Without historical data, any evidence of pillowing would trigger this inspection. These inspections could be performed concurrently using a single scanner such as the Tektrend UNIT or the Boeing MAUS IV. Inspections for cracks can be achieved simultaneously with another EC scan optimised for this application. Figure 10.1 shows a flowchart describing such inspection process.

The fusion of NDT data described in the previous paragraph can fully characterise the damage state of the lap joint structure. Before data fusion algorithms can be applied to individual NDT results, registration of the data must be performed. Fortunately, the lap joint construction is essentially two-dimensional; that is, curvature of the joint can be neglected. This allows the use of simple matrix transformations to rotate and scale different NDT results onto the same co-ordinate system. Many of the disparate NDT methods considered here could also be performed simultaneously using one scanning device, which would simplify registration.

For the metric of pillowing, the maximum pillowing in the area between adjacent rivets can be estimated using enhanced visual (EV) NDT methods. This measure can then be used to estimate the average amount of material converted to corrosion in this area. The maximum pillowing measure may also be used in estimating the pillowing stresses that can be used as input for damage tolerance calculations.

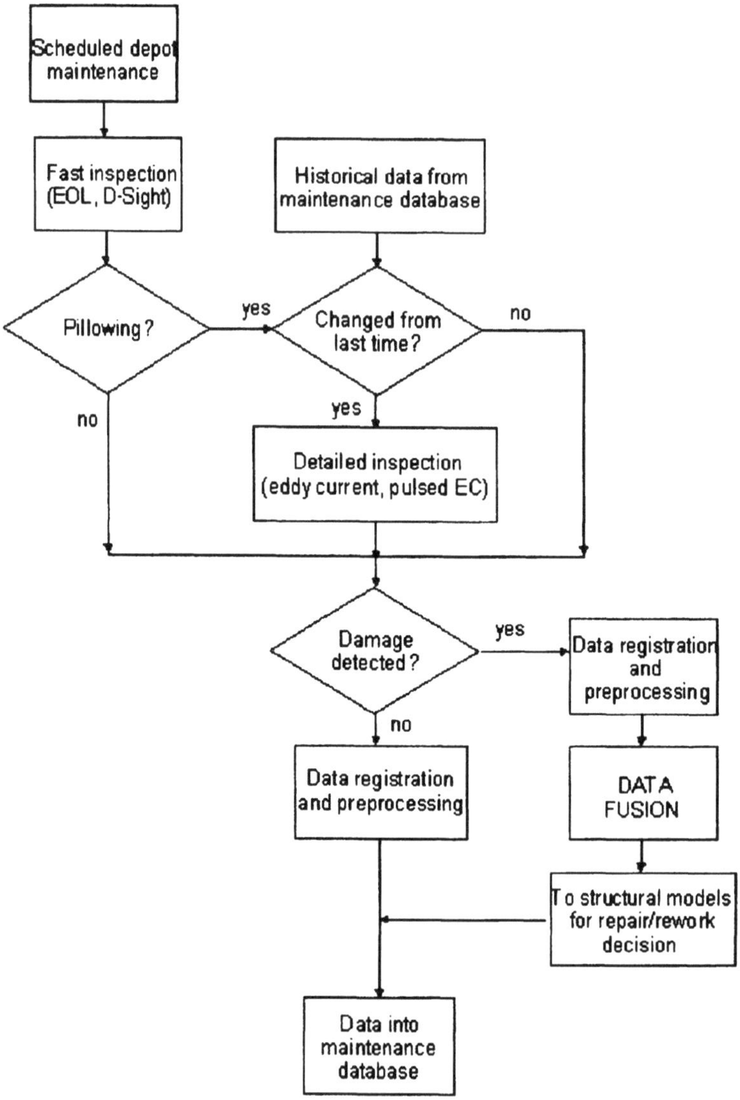

Fig. 10.1 Schematic illustration of the role of data fusion in aircraft maintenance

Previous work demonstrated manual and automated pillowing estimation from EOL and D-Sight EV methods [13-15]. For manual estimation, calibration images are used. The inspector matches the test results with calibration images of known pillowing deflections. Algorithms have been developed to perform this step automatically. Finite element models of the individual joint construction are then used to transform the maximum pillowing deflection to the amount of material lost in the area between adjacent rivets, and to estimate pillowing stresses [16]. Therefore, the resolution of the processed EV results for corrosion is the area between adjacent rivets.

The metric of thickness loss by layer is a difficult requirement. Total material converted to corrosion can be estimated from pillowing, but the layer on which the corrosion occurs cannot be determined from the EV results. The thickness of the first layer in a thin, multi-layer system such as a lap joint can be estimated with reasonable accuracy using EC or ultrasonic testing (UT) techniques. Pulsed EC techniques currently developed by a number of organisations show promise for the quantification of corrosion per layer, but have not yet been proven ready for depot-level use.

At first glance, one may assume that a simple subtraction of the first layer thickness measured by EC or UT from the total material loss inferred by pillowing measurements would yield the required data on the individual layers. Unfortunately, due to the tolerance in the as-manufactured thickness of the aluminium sheet used in the construction of many lap joints, this information is not yet sufficient for input into a fracture mechanics type model. The tolerance varies slightly depending on sheet thickness, but is on the order of 5% in the thickness typical in lap joint construction.

It is necessary to determine what thickness loss has been caused by corrosion, as the pitting and increased surface roughness cause by corrosion reduces the time to crack initiation, and speeds crack growth. This means that the life of a joint which is 2% thinner than the nominal will be much shorter if the 2% loss is due to corrosion than if the 2% loss is due to sheet tolerance [5].

There are two main methods of determining the cause of thickness loss from NDI results. The first is by analysis of the pattern of thickness loss measured using NDT techniques. This applies especially to EC or UT techniques, which produce relatively high-resolution maps of thickness. Changes in thickness due to corrosion tend to occur at a higher spatial frequency than changes due to as-manufactured variance. Small areas with large changes in thickness are readily identified as being due to corrosion.

The second method of determining the cause of thickness loss is to use pillowing measurements. Ideally, at the time of manufacture, the joint will have no deformation at all. Any subsequent pillowing-type deformation would then be the result of material being converted to corrosion product. In reality, there are a number of other effects, which deform the surface of a lap joint, including the installation of rivets, and mechanical loading. These can be distinguished from corrosion-induced pillowing by the shape of the deformations, but at small levels of corrosion this is an uncertain process. This uncertainty can be almost entirely mitigated by using a baseline. If the existing deformations are measured at time of manufacture, and repeatedly during the aircraft lifetime, changes in deformation can be analysed to quantify corrosion. This is a relatively simple task given the sophisticated maintenance databases in use today, and has been adopted by fleet managers in the Canadian Forces as well as the Royal Air Force.

Another key metric required for some of the new models of corrosion/fatigue is the topography of the corroded surface [17]. While it is likely to prove impractical to directly measure the roughness of the faying surfaces, the authors believe that metallurgical studies may be able to define the correlation between actual material loss and typical corrosion topographies. This metric also requires identifying whether corrosion or manufacturing variability is the cause of thickness changes.

10.4 APPLICATION OF NDT DATA FUSION

Two sections from a lap joint specimen were inspected as an example to perform data fusion. This specimen was cut out of a larger lap joint specimen from the IAR Specimen Library. This lap joint originated from a Boeing model 747-151. The aircraft was first flown 17 June 1970, and withdrawn from use 20 June 1994. It was mostly used for long haul flights on Pacific routes. The final flight hours accrued by the aircraft were 61,318 and the final flight cycles were 16,612. The specimens were removed 14 October 1994. The location of the specimen was stringer (STR) 34L, body station 1900-1950. The length of the lap joint is 106.68 cm long. This specimen is part of the IAR Specimen Library, number 185B. There are 34 columns of rivets in this specimen. The lap joint is constructed out of 2024-T3 aluminium, with 1.6 mm thick top and bottom layers.

10.4.1 Teardown

The lap joint was cut into four sections and was carefully disassembled. To disassemble lap joint specimens, first the shop heads of the rivets in any stringers are drilled and chiselled off and the stringers are removed. Next, the rear surface of the second layer skin is imaged to record any corrosion damage between the stringers and skin. Then the rivets are removed following the process described below. The two layers are separated using a thin wedge between the skins to defeat any bond. The faying surfaces are imaged to record the location of corrosion damage and presence of corrosion product.

The procedure used for rivet removal follows. All rivets are drilled undersize to a depth that releases most of the interference/compressive force caused during the bucking of the rivets in the holes, but are not drilled completely through the length of the rivet. The joint is positioned face down on a rigid plate having a hole just larger than the rivet head and a pin probe attached to an arbour press is used to manually force the rivet out of the joint.

The corrosion damage was evident on the second layer outboard surface. The faying surface epoxy adhesive layer remained intact and completely bonded to the inboard surface of the first layer skin and thus there was little corrosion damage to the first layer skin. The skins from section 1 and 3 are shown immediately after disassembly and cleaning in figures 10.2 and 10.3 respectively. These sections are used as example specimens.

The film adhesive used in these specimens is an epoxy adhesive with a carrier cloth. To remove the adhesive a judicious use of mechanical abrasion through the epoxy and carrier cloth is required to reduce the amount of adhesive on the surface without contacting the surface. By reducing the adhesive layer to a minimum the epoxy could be softened and its bond broken with a commercial paint stripper.

The remaining adhesive was then softened and removed by applying a commercial paint stripper that left the original primer coating. Finally a more aggressive commercial paint stripper was used to chemically remove the exterior paint system and primer coat from the first and second layer skins.

Corrosion products are removed in two steps. First, fully converted products are carefully removed by mechanical means without damaging the remaining base material. Then the layer of the surface which is in transition between the metal and oxide is removed by a chemical solution in an ultrasonic bath. This is done in several sessions, with visual examinations of the surfaces between each session to ensure that only the corrosion layer is removed.

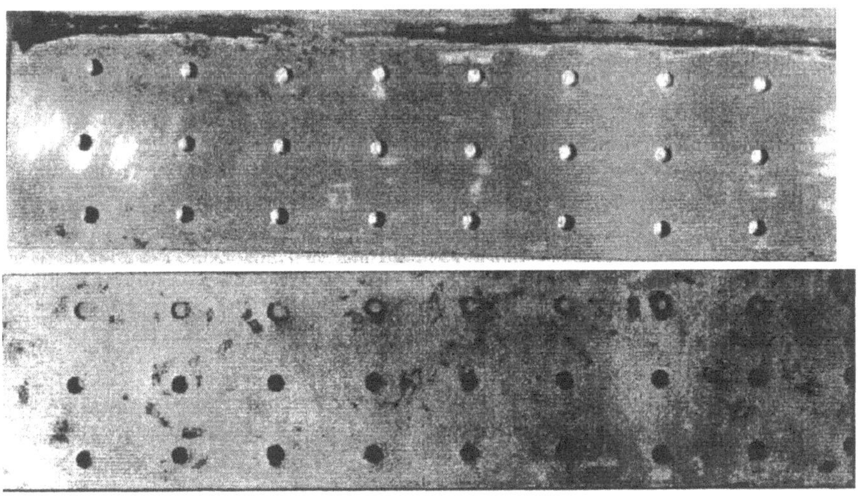

Fig. 10.2 Faying surfaces of section 1 of a lap joint, after disassembly and cleaning

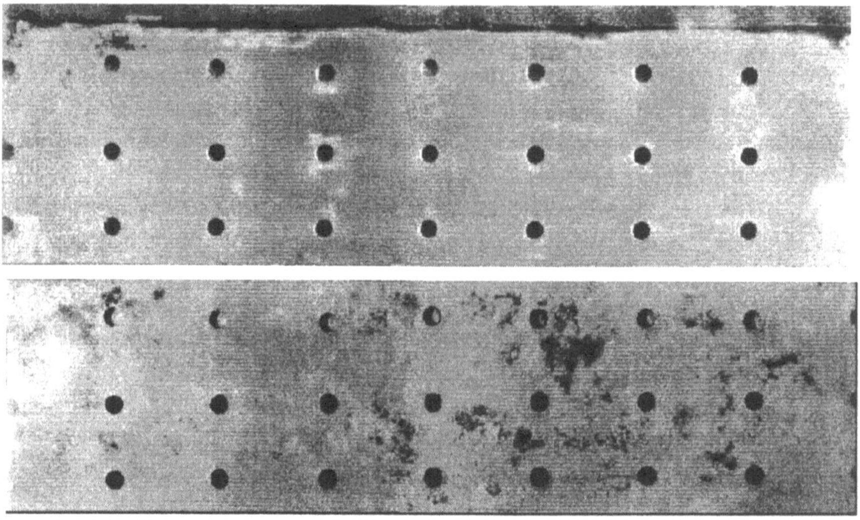

Fig. 10.3 Faying surfaces of section 3 of a lap joint, after disassembly and cleaning

10.4.2 Verification

A method has been developed at the IAR for measurement of the thickness of solid structures using X-radiography [18]. The transmission of X-rays through a solid body of homogeneous material is a function of the thickness of the material. To convert from the transmitted intensity to the thickness, the part to be measured is placed next to a wedge-shaped calibration piece of the same material. The pieces are exposed, and the results recorded on film. The film is digitised, and the IAR's NDI analysis software is used to generate the relationship between intensity and part thickness using the calibration wedge. Once this relationship is obtained, it can be used to calculate the thickness of the unknown specimen. The specimens are not always exactly made of the same alloy as the calibration piece, and mechanical measurements on the unknown specimen are used to confirm the accuracy of the radiograph measurement and provide a correction factor if required.

Calibrated radiographs for the individual layers of the specimen are shown in figure 10.4 for section 1 of the specimen and figure 10.5 for section 3 of the specimen. The results of the destructive examination of the specimen, averaged over the area between rivets to match the inspection results, are shown in figure 10.6. The total thickness of the joint, averaged over the area between rivets, was calculated and is shown in figure 10.7.

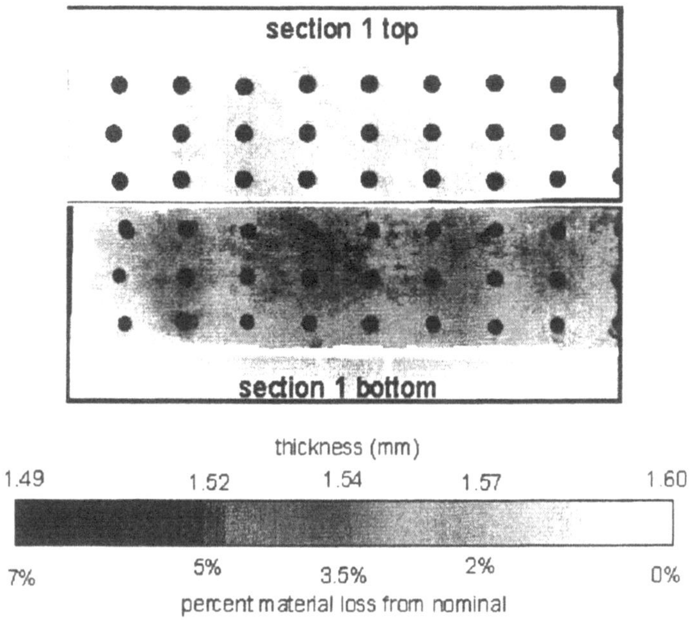

Fig. 10.4 Radiographs of section 1 of the specimen

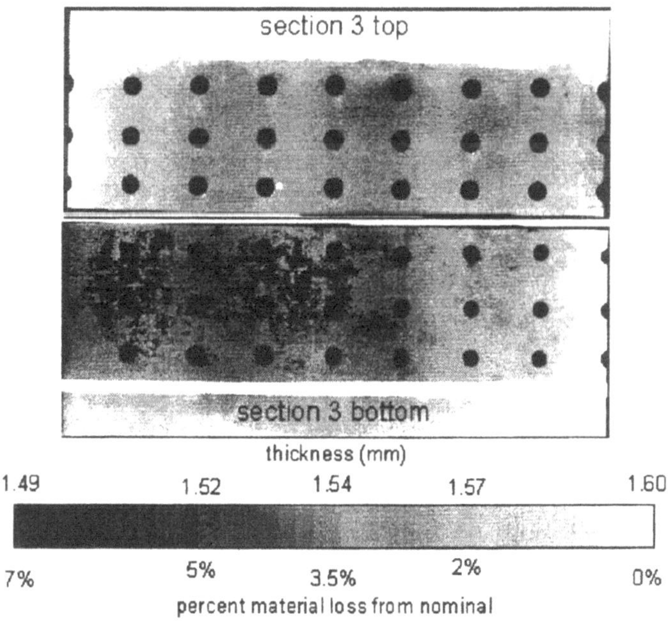

Fig. 10.5 Radiographs of section 3 of the specimen

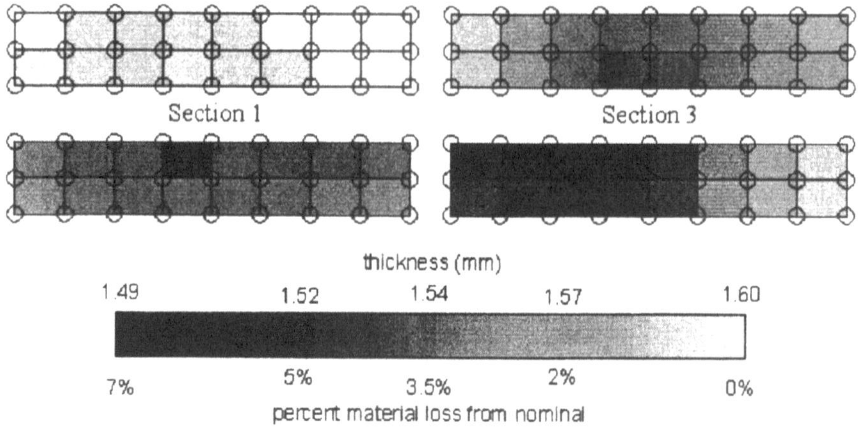

Fig. 10.6 Thickness of individual layers in the joint, from radiographs, averaged over areas between adjacent rivets

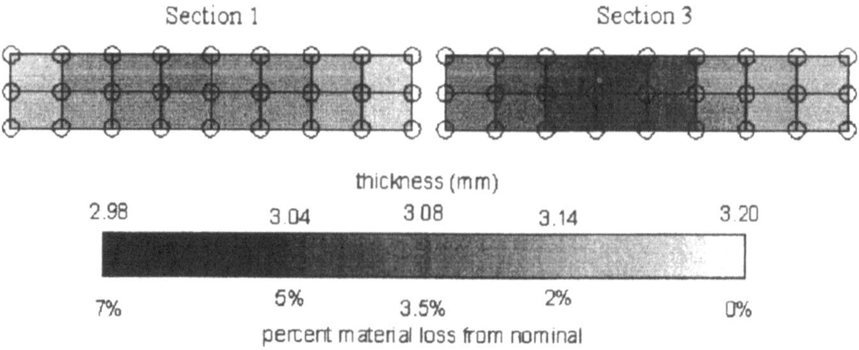

Fig. 10.7 Combined thickness of both layers in the joint, from radiographs, averaged over areas between adjacent rivets

10.4.3 NDT Results

The resolution of NDT techniques such as eddy current or ultrasound commonly used for lap joint inspection is much higher than that of processed EV techniques. Due to the limitations of the EV techniques, results of individual NDT techniques have been presented averaged over the area between adjacent rivets. While there is a loss of resolution, it must be remembered that repair/rework decisions on an aircraft are currently made on a frame to frame (or typical body station length of about 500 mm) consideration. Ultimately, the new damage tolerance/durability assessment models will dictate the required resolution.

The original NDT results are presented below for a number of different techniques. The results are accompanied by registered and processed data derived from the individual results. Where the processed results are given in terms of material thickness, the palette used matches the palettes used in the previous section for the radiograph-measured thickness. Due to the uncertainty in measuring such small thickness changes, the results are placed into three categories; no discernible loss, 0 to 2% thickness decrease from nominal, and 2 to 5% thickness decrease.

Eddy Current Measurements

Eddy current measurements can be used with reasonable accuracy to determine the top layer thickness in a lap joint. In this work, a Zetec MIZ40 eddy current instrument was used with a probe developed by the IAR and RDTech of Quebec City specifically for this purpose. Measurements were performed simultaneously at four frequencies. In order to measure material loss on the top layer only data obtained at 12kHz was analysed. The original eddy current results and the interpretation, averaged over the area between adjacent rivets, are shown in figure 10.8.

With this specific technique, the amplitude of the imaginary portion of the eddy current signal is used to indicate changes in thickness. The probe is balanced on a calibration piece of the same thickness as the uncorroded specimen. A reduction in the thickness of the specimen being inspected results in an increase in the amplitude of the imaginary component of the eddy current signal. The amplitude of the signal has been colour-coded, with white areas representing larger amplitude or greater material loss. The areas of material thinning shown in figure 10.8 were interpreted as being due to corrosion. This is indicated by irregular shapes and a lack of low spatial frequency trends.

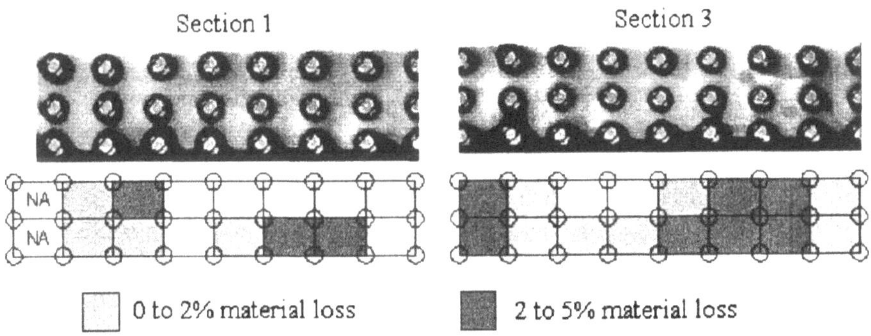

Fig. 10.8 Eddy current results and interpretation for the top layer

Pulsed Eddy Current Measurements

A consortium including IAR, Canada's Department of National Defence, and Tektrend International of Montreal developed the pulsed eddy current system used in this work. The interpretation of the results is still under development, however it has been shown to be sensitive to small amounts of material loss, and have potential for quantification and location of flaws [19-21].

A pulsed eddy current inspection of specimen was performed, and the results are shown as a lift-off point of intersection (LOI) scan in figure 10.9 [22]. Pulsed EC test results provide a voltage-time curve for each point, similar to a full wave ultrasonic inspection. Generally, the measured signal is subtracted from a reference signal defined on a location with no flaw. Once this operation is done, a number of features of the resulting signal can be used to locate flaws in depth, and estimate their size.

In the data presented in figure 10.9, lighter colours correspond to low material thickness. The results do not indicate on which layer the material thinning is located. The areas of material thinning shown in the interpretation of the pulsed EC test were due to corrosion only, as indicated by the irregular shapes and a lack of low spatial frequency trends.

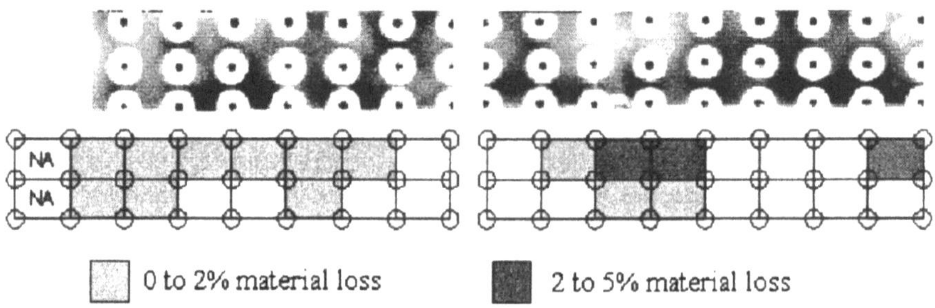

Fig. 10.9 Pulsed eddy current results and interpretation averaged between rivets for the top layer

D-Sight Measurements

As previously discussed, it has been shown that the D-Sight inspection technique can be used to quantify pillowing and associated material loss in lap splice joints. For this specimen configuration, the current state of the technique requires an operator to match inspection results to calibration images. A set of four calibration images was used, each representing D-Sight images for different known levels of pillowing. The inspections from the specimen were marked to show areas which matched levels of pillowing from the calibration images.

The pillowing measurement is for the maximum deformation between a group of four neighbouring rivets. It assumes even corrosion loss in this area. Based on previously reported work, this pillowing deformation can be used to calculate the amount of material converted to corrosion product, and therefore the total thickness loss at the faying surfaces [14].

The D-Sight inspections of the specimen and associated interpretations are shown in figure 10.10. It should be noted that the D-Sight images are not processed. The sensitivity to pillowing is maximum at the left side of the image, and reduces gradually towards the right side. This can be corrected for if visual interpretation is required.

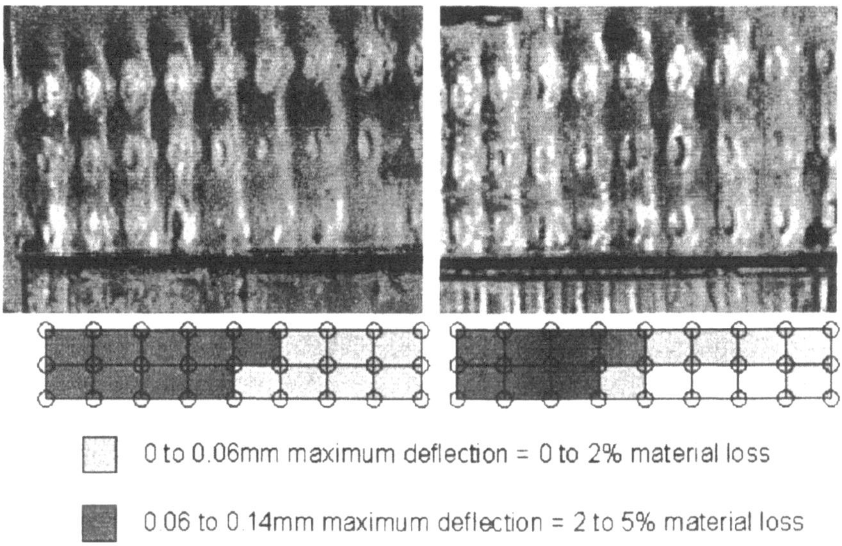

Fig. 10.10 D-Sight images of the specimen and associated interpretations

Edge-of-Light Measurements

It has been shown that the EOL technique can be used to detect and quantify pillowing, and thus total thickness loss, in a lap joint [15]. Again, this technique can be automated for particular joint configurations, but a manual examination was performed here using comparisons to EOL inspections of known pillowing. The EOL instrument used for this work is a hand-held experimental unit, and it is believed that the technique is capable of higher performance than demonstrated herein. Prototypes are being designed for typical NDI applications.

The original EOL inspection results of the specimen are shown in figure 10.11, with a map of maximum pillowing and the associated map of total joint thickness loss.

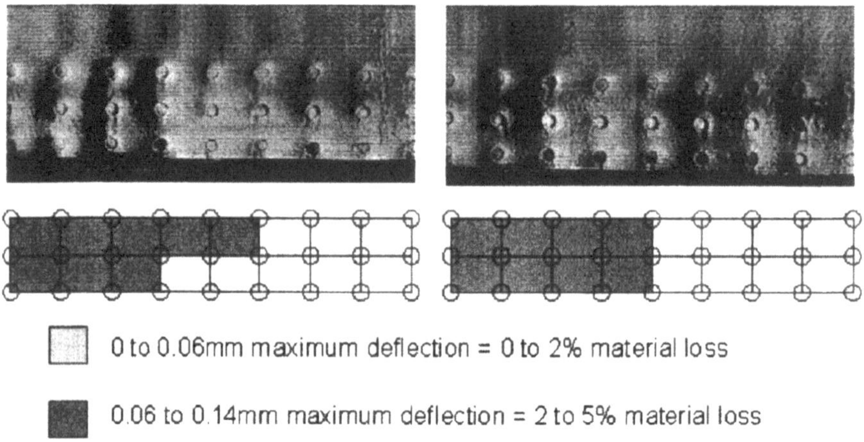

Fig. 10.11 EOL inspections of the specimen and associated interpretations

10.4.4 Discussion of NDT Results

Both D-Sight and EOL detected pillowing deformation in the specimen were interpreted manually. The D-Sight image showed more deformation than the EOL image, which may be due to other sources of deformation than corrosion pillowing. Both techniques measured high levels of pillowing in the area around rivet columns 1 to 4, which was corroded, but not as badly as other areas. Both scans picked up the most corroded areas, but D-Sight was more sensitive to lower levels of corrosion which exist throughout this specimen.

Conventional eddy current measurements of the first layer thickness compared reasonably well with destructive tests. These measurements can be confounded with additional effects, such as changes in lift-off due to pillowed surface or to changing gap between the layers. It is hoped that the pulsed eddy currents technique currently under development will prove more accurate and reliable than the conventional eddy current. The results presented here are proportional to the total thickness loss in the joint; research is still underway to determine thickness loss per layer estimates from pulsed EC data.

10.4.5 Fusion of NDT results

In the idealised maintenance situation presented in figure 10.1, historical NDT data would be fused with the newly acquired data to determine if any changes in specimen condition had occurred. This would be done by comparing pillowing measurements, which can be made using a number of different EV NDT techniques. Changes in pillowing would be used to call out detailed inspections using EC or pulsed EC techniques. In the example presented herein, historical data was not available, and the EV results are fused with EC and pulsed EC test results in order to provide estimates for all the metrics believed necessary for structural models of the lap joint structure.

Given the measurements of pillowing using D-Sight, a map of the total material loss can be generated. With the first layer thickness estimates from EC and the total thickness estimates, the second layer thickness can be estimated. Results of this data fusion process are shown below compared with radiographic results. The total joint results are shown in figure **10.**12, and layers one and two are shown in figures 10.13 and 10.14 respectively.

The data fusion method employed here is rather simplistic, but nevertheless provides improvements over individual NDT data. The results are available in quantitative terms, in terms of pillowing and thickness. Results from the individual techniques can be directly compared. Any additional results can be processed onto the same co-ordinate system, and if required, other data fusion algorithms can be employed. For example, the uncertainty in determining the cause of thickness loss may be reduced by using techniques such as Bayesian inference or Dempster-Shafer evidential reasoning. The efficiency of these mathematical approaches has been demonstrated on NDT data [11], although most of the published applications of data fusion to NDT are of moderate complexity and involve pixel or feature level fusion [23-27].

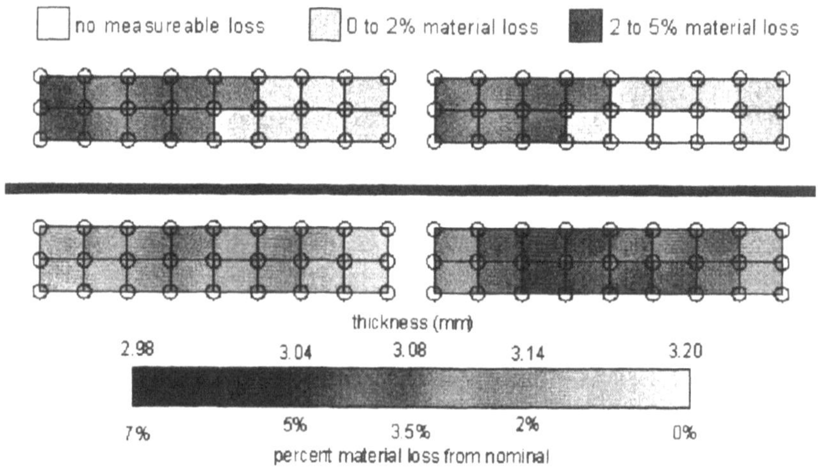

Fig. 10.12 Comparison of total joint thickness loss estimates between NDT results (top) and disassembly observations (bottom)

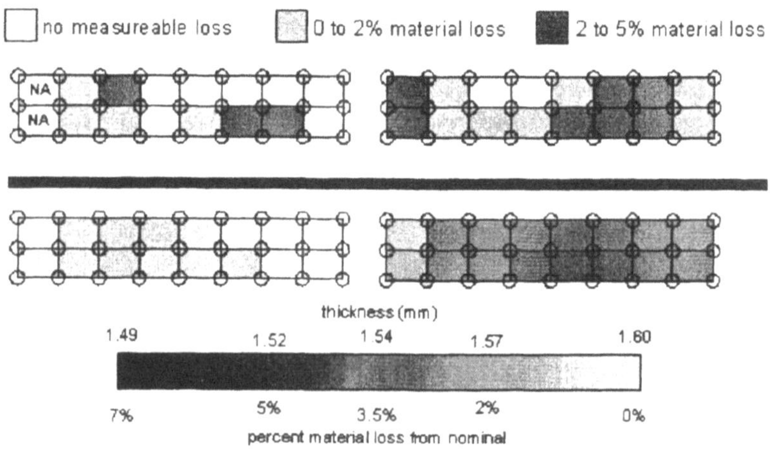

Fig. 10.13 Comparison of first layer thickness estimates between NDT results (top) and disassembly observations (bottom)

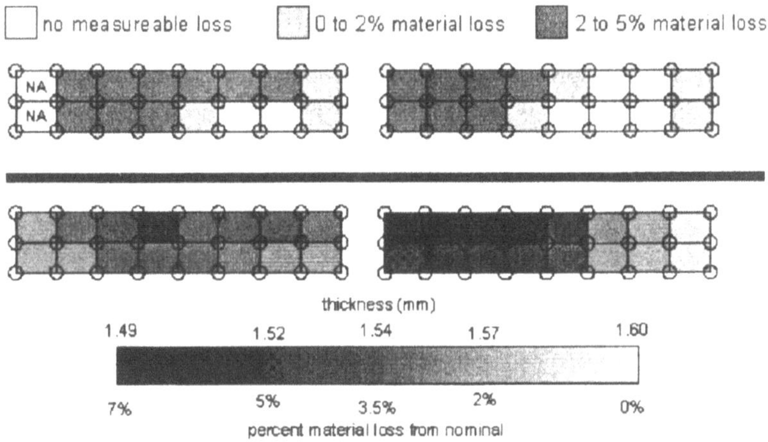

Fig. 10.14 Comparison of second layer thickness estimates between NDT results (top) and disassembly observations (bottom)

10.5 SUMMARY

The complex shapes and multiple layer structures which exist in transport aircraft are subject to a variety of different damage due to corrosion. The fuselage lap joint is a common element of construction, and is also a difficult problem for NDT. This problem can be successfully addressed using data fusion techniques on information gathered with commercially available NDT instruments.

It has been demonstrated in this work that several inspections are required in order to fully characterise the condition of a lap joint structure in terms of layer thickness, material loss due to corrosion, pillowing deformation and cracking. For example, due to sheet tolerances, both the presence of corrosion pillowing and material thinning are required to confirm the existence of corrosion. In the case of cracking, the eddy current techniques used to detect small cracks are almost opposite to the EC techniques to detect corrosion.

The required inspections and signal interpretation can be performed manually by qualified inspectors, but for practical reasons this will be impossible in a fleet maintenance situation. Indeed, the K/C-135 contains some 900 feet of lap splice joints and 1000 square feet of doublers, which are also possible sites for hidden corrosion. The inspection of all these areas on a fleet of more than 500 K/C-135 aircraft operated by the USAF will generate an enormous amount of data. Practical use of this data will require automation of data handling and interpretation, as well as integration of inspection data and results with maintenance databases. These items provide all the prerequisite for data fusion.

Other aircraft structures affected by corrosion and other damage modes such as rivet failure have not been directly addressed here. Newer aircraft often have composite materials joined to metallic materials, each with very different inspection requirements. It is believed that data fusion techniques could bring together results from various NDT techniques such as thermography, shearography, UT, EC and whatever other methods may be used. The end result would be to provide the inspector with data that is easier to view, manipulate and interpret, thus leading to improved decisions and safety.

REFERENCES

1. Hoeppner D.W., Grimes L., Hoeppner A., Ledesma J., Mills T., Shah A., *Corrosion and fretting as critical aviation safety issues: case studies, facts, and figures*, Proc. 18th Symp. of the Inter. Committee on Aeronautical Fatigue, 1995, Melbourne, Australia, 1:87-106.
2. Cole G.K., Clark G., Sharp P.K., *Implications of corrosion with respect to aircraft structural integrity*, Airframes and Engines Division, Aeronautical and Maritime Research Lab., Australia, Report N°DSTO-RR-0102, 1997.
3. Komorowski J.P., Bellinger N.C., Gould R.W., *The role of corrosion pillowing in NDI and in the structural integrity of fuselage joints*, Proc. 19th Symp. of the Inter. Committee on Aeronautical Fatigue, Edinburgh, Scotland, 1997, 1:251-266.
4. Bellinger N.C., Komorowski J.P., Gould R.W., *Damage tolerance implications of corrosion pillowing on fuselage lap joints*, J. of Aircraft, 1998, 35(3):487-491.
5. Eastaugh G.F., Merati A.A., Simpson D.L., Straznicky P.V., Krizan D.V., *The effects of corrosion on the durability and damage tolerance characteristics of longitudinal fuselage skin splices*, Proc. USAF Aircraft Structural Integrity Conf., 1998, San Antonio, USA.
6. Wallace W., Hoeppner D.W., Kandachar P.V., *Aircraft Corrosion: Causes and Case Histories*, AGARD Corrosion Handbook, AGARD-AG-278, 1985, 1.
7. Brooks C.L., Prost-Domasky S., Honeycutt K., *Corrosion is a structural and economic problem: transforming metrics to a life prediction method*, NATO RTO's Workshop 2 on Fatigue in the Presence of Corrosion, 1998, Corfu, Greece.
8. Bellinger N.C., Komorowski J.P., *Environmentally assisted cracks in 2024-T3 fuselage lap joints*, Proc. 3rd Joint FAA/DoD/NASA Conf. on Aging Aircraft, 1999, Albuquerque, USA.
9. Bellinger N.C., Gould R.W., Komorowski J.P., *Repair Issues for Corroded Fuselage Lap Joints*, Proc. World Aviation Congress, 1999, San Francisco, USA.
10. Forsyth D.S., Chapman C.E., Lepine B.A., Giguiere S., *Nondestructive inspections of K/C-135 "flyswatter" fitting*, National Research Council Canada, 2000, Report N°SMPL-LTR-2000-0241.
11. Forsyth D.S., Komorowski J.P., *The role of data fusion in NDE for aging aircraft*, , Proc. SPIE Nondestructive Evaluation of Aging Aircraft, Airports, and Aerospace Hardware IV, 2000, 3994:47-58.
12. Gros X.E., *NDT Data Fusion*, Butterworth-Heinemann, 1997.
13. Alcott J. *et al.*, *Results of on-aircraft demonstration of nondestructive inspection equipment to detect hidden corrosion on USAF aircraft*, 1994, ARINC Report N°F41608-90-D-0544-SD01-04.
14. Forsyth D.S., Komorowski J.P., Gould R.W., *the use of solid film highlighter in automation of D sight image interpretation*, Proc. SPIE Nondestructive Evaluation of Aging Aircraft, Airports, and Aerospace Hardware II, 1998, 3397:50-56.
15. Komorowski J.P., Bellinger N.C., Gould R.W., Marincak A., Reynolds R., *Quantification of corrosion in aircraft structures with double pass retroreflection*, 1996, Canadian Aeronautics and Space J., 42(2):76-82.
16. Forsyth D.S., Gould R.W., Komorowski J.P., *Correlation of enhanced visual inspection image features with corrosion loss measurements*, Proc. 3rd Inter. Workshop on Advances in Signal Processing for NDE of Materials, 1998, 3:365-372.

17. Bellinger N.C., Krishnakumar S., Komorowski J.P., *Modelling of pillowing due to corrosion in fuselage lap joints*, Canadian Aeronautics and Space J., 1994, 40(3):125-130.
18. Brooks C., Peller D., Honeycutt K.T., Prost-Domasky S., *Predictive modeling for corrosion management: modeling fundamentals*, Proc. 3rd Joint FAA/DoD/NASA Conf. on Aging Aircraft, Albuquerque, USA, 1999.
19. Chapman C.E., Marincak A., *Corrosion Detection and Thickness Mapping of Aging Aircraft Lap Joint Specimens Using Conventional Radiographic Techniques and Digital Imaging*, National Research Council Canada, Report N°NRC-LTR-ST-2045, 1996.
20. Bieber J.A., Tai C., Moulder J.C., *Quantitative assessment of corrosion in aircraft structures using scanning pulsed eddy current*, Review of Progress in QNDE, 1998, San Diego, USA, 17:315-322.
21. Burke S.K., Hugo G.R., Harrison, D.J., *Transient eddy current NDE for hidden corrosion in multilayer structures*, Review of Progress in QNDE,1998, San Diego, USA, 17:307-314.
22. Lepine B.A., Wallace B.P., Forsyth D.S., Wyglinsky A., *Pulsed eddy current method developments for hidden corrosion detection in aircraft structures*, CSNDT J., 1999, 20(6):6-15.
23. Giguère S., Dubois S., *Pulsed eddy current: finding corrosion independently of transducer lift-off*, Review of Progress in QNDE, 2000, Montréal, Canada, 19A:449-456.
24. Bossi. R.H., Nelson J., *Data fusion for process monitoring and NDE*, Proc. SPIE Nondestructive Evaluation for Process Control in Manufacturing, 1996, 2948:62-71.
25. Gros X.E., Bousigue J., Takahashi K., *NDT Data Fusion at pixel level*, NDT&E Inter., 1999, 32(5)283-292.
26. Yim J., Udpa S.S., Udpa L., Mina M., Lord W., *Neural network approaches to data fusion*, Review of Progress in QNDE, 1995, Snowmass Village, USA, 14:819-826.
27. Forsyth D.S., Komorowski J.P., *Development of NDT sensor/data fusion in support of ageing aircraft requirements*, Proc. of the ASNT 8th Annual Research Symp., 1999, Orlando, USA.

11 MEDICAL APPLICATIONS OF NDT DATA FUSION

Pierre JANNIN, Christophe GROVA,
Bernard GIBAUD
Université de Rennes
Rennes, France

11.1 INTRODUCTION

The fusion of different sources of information has always been a component of medical practice. The complexity of biological phenomena is such that they cannot be explained with a single exploration. The complementary nature of the available exploration techniques (modalities) helps the physician in refining his diagnosis, preparing or performing therapeutic procedures. In this respect it is interesting to note that the development of new medical modalities has not led to the replacement of former ones, and that obviously there is no single modality providing the clinician with all possible sources of information. Before the development of computerised registration tools, data fusion involved pure mental matching of the data sets based on well-known common structures, the matching of the rest of the data being mentally interpolated from this initial step. The parallel emergence of new digital medical imaging devices, communication networks and powerful workstations has made it possible not only to display images but also to transfer and process them. Image processing methods have recently been developed to perform direct, automatic data matching. These methods define the multimodal data fusion topic. They have modified the way multimodal matching is performed as well as the way multimodal information is used; moving from a mental to a computer assisted fusion process. This evolution has led to a more accurate, more visual, more quantitative and therefore more objective fusion process. These data fusion capabilities have contributed to the development of advanced clinical procedures such as image-guided therapy. In this chapter we present the current state of the art of data fusion methods for medical applications, with an emphasis on development and validation. Therefore this chapter focuses more on the influence of the application context than on the methods themselves. An exhaustive list of papers describing methods or applications is not included as previous surveys exist with an extensive list of references: Maintz *et al.* [1] (283 references), Maurer *et al.* [2] (202 references), Van Den Elsen *et al.* [3] (125 references), Hawkes [4] (51 references), Brown [5] (101 references). However, we have endeavoured to extract the major trends representative of current approaches in medical data fusion from the literature and from our own experience.

The chapter is organised as follows. First, the medical application context of data fusion is presented. The available multimodal data, as well as definitions and clinical applications are introduced. The data fusion methods used for matching multimodal data sets are presented, and the fact that data fusion cannot be limited to registration is emphasised. Pre-processing tools are often required before registration itself and notably application tools need to be developed to exploit the results of registration: visualisation, interaction and analysis. Validation is a major issue for the clinical use of data fusion. Next, different facets of validation and the ways to implement it are described. A detailed clinical application illustrates the implementation and validation of data fusion methods. The discussion section provides indications to help choose an adequate data fusion method and attempts to explain why industrial data fusion products are not widespread. Finally the "Prospective" section lists research trends for the coming years.

11.2 MULTIMODAL MEDICAL IMAGING DATA FUSION

11.2.1 The Diversity of Medical Multimodal Data

Various types of data may be involved in medical applications of data fusion such as measures (e.g. images, signals), location (e.g. patient, surgical tools, robot, imaging devices) and *a priori* knowledge (e.g. clinical case database, anatomical or physiological models). Two main categories of measures can be distinguished: anatomical and functional data. Anatomical modalities provide morphological representations of specific areas of the anatomy (e.g. bone, brain, soft tissues, heart) based on their physical properties such as density of the tissues (e.g. X-ray radiography, computed tomography (CT), portal imagery), hydrogen proton density (e.g. magnetic resonance imaging (MRI)) or tissue acoustic properties (e.g. echography). Contrast agents can be used to enhance specific parts of the anatomy, particularly vessels. These are magnetic resonance angiography (MRA), X-ray angiography, digital subtraction angiography (DSA). Functional modalities provide information related to a specific function of the living body. Some of these (e.g. functional MRI (fMRI), MR and CT perfusion, single photon emission computed tomography (SPECT), positron emission tomography (PET)) explore a specific metabolic process (e.g. brain or heart perfusion, glucose consumption), others such as electro-cardiography (ECG), electro-myography (EMG), electro-encephalography (EEG), magneto-encephalography (MEG), electro-corticography (EcoG), stereo-electro-encephalography (SEEG) explore bio-electric or bio-magnetic phenomena (e.g. neuronal electrical activity related to a specific stimulation). In SPECT and PET, properties of a metabolic process are highlighted using specific products or molecules (e.g. water, glucose) of which effects are monitored in the body with the help of radioactive tracers such as technetium 99 or iodine 131. All these modalities have various spatial, temporal and contrast resolutions (figure 11.1). This brief overview is not exhaustive, but shows the complementary aspect and diversity of modalities encountered.

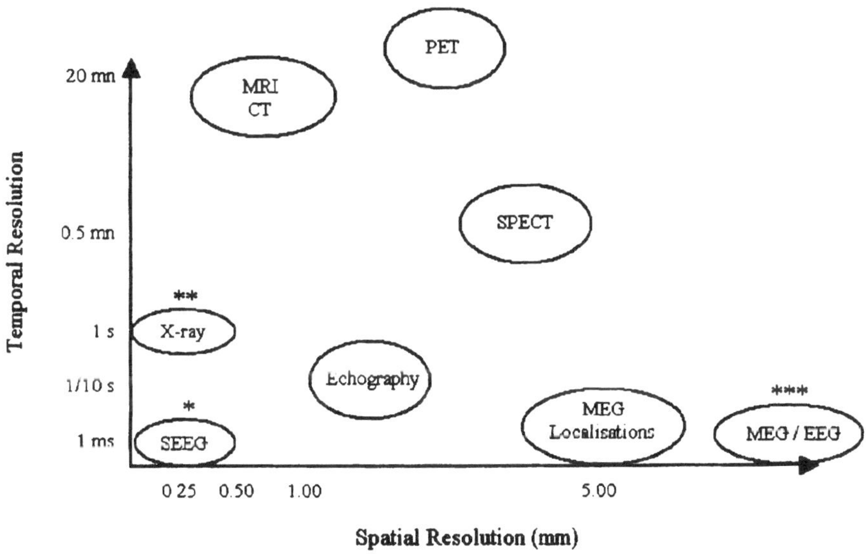

Fig. 11.1 Temporal (milliseconds, seconds, minutes) *vs.* spatial resolution of some medical modalities; * *in situ* measurement, ** projection, *** surface measurement

11.2.2 Rationale of Medical Data Fusion

When a physician or surgeon decides on a multimodal imaging approach for a patient there is always a clinical objective in mind. This clinical objective corresponds to clinical goals to be reached through the exploitation of additional information resulting from a data fusion process: e.g. overlapping areas, distance between areas, comparison of statistical features inferred from measurements performed on regions of interest. Obviously clinical objectives are adjusted according to the clinical procedure (e.g. diagnosis, planning or therapeutic procedure), to the anatomical area concerned by the clinical procedure and to the patient' specific pathology. The use of data fusion in medicine requires an accurate definition of clinical objectives. The clinical context of a data fusion process can be defined as the set of circumstances, of a clinical nature, in which a fusion process takes place; including the clinical objective and the choice of imaging modalities to explore an anatomical area of the patient. For instance, a clinical context of a data fusion process can be to compare the location of eloquent brain areas detected with functional imaging, with the location of a lesion visible in anatomical imaging, in order to plan a neurosurgical procedure.

In general, data fusion aims at integrating multiple sources of information to support a decision. In the medical field, information can be composed of images and signals but also of clinical observations, interviews, laboratory test results, videos, individual clinical history, etc. In this chapter the scope of data fusion is restricted to the matching of various images or measures of physical or physiological phenomena concerning one or more physical entities considered as significantly similar by the application end user. "Significantly similar" means that the different measures correspond to the same anatomical region or that both measures conform to a common *a priori* model. This distinction is justified by the case of fusion between data sets coming from different subjects or patients. In this case the anatomical structures are not exactly the same but it is assumed that invariant (similar) shapes or information exist in both data sets. This is notably relevant in neuro-imaging, for studying the anatomo-functional variability among a population. In this case registration between different MRI data sets of different subjects assumes that the overall shapes of the individual brains are similar in spite of local anatomical dissimilarities. The data fusion issue is due to dissimilarities between the data to be matched, arising from different acquisition conditions and/or different types of measurands. These dissimilarities partly correspond to the relevant information one wants to highlight, using data fusion methods (e.g. complementary nature of measures) and partly to variations in the measurands for which we want to compensate using registration methods [5]. Variations can be geometric or intensity related, but generally attempts are made to compensate for geometric variations. A data fusion method provides a physician or surgeon with a set of tools for matching and analysing multimodal data sets. A registration method is part of a data fusion method used to compute a transformation between multimodal data sets in order to compensate for geometric or intensity related variations. These multimodal data sets explore a common physical entity, which is generally an anatomical area. Therefore, the anatomical substrate is often used as a common reference to match multimodal functional data. The basic principle of registration methods is to use common information from both data sets to compute the transformation. The fusion context of a data fusion process consists of a set of circumstances of a data fusion process, focused on the data to be matched: intrinsic characteristics (e.g. dimension, modalities) and assumptions related to similarities or dissimilarities between these data sets (e.g. same anatomical area, moving structures, movement during acquisition, *a priori* knowledge). A medical application of data fusion is the complete specification of a data fusion process according to a given clinical context and a fusion context.

11.2.3 Medical Applications of Data Fusion

Medical applications of data fusion may be classified according to the fusion context. Four fusion contexts can be distinguished. Characteristics of each context and examples of corresponding clinical contexts as well as medical applications are described next.

Fusion Context I: Intra patient and intra modality registration

It is the registration between different data sets of a single modality concerning a single patient. Three main clinical objectives may correspond to this fusion context that include: monitoring of changes in patient anatomy over time, comparing different patient conditions, and subtraction imaging. Monitoring changes in patient anatomy over time allows growth monitoring or the capability to study the evolution of a lesion (e.g. tumour, multiple sclerosis) or the effects of a treatment. In the context of surgery or radiotherapy it may allow the comparison of pre and post-treatment images. The second objective is used to compare different patient conditions. Activation measurements with functional MRI are performed by statistical comparison of images acquired during a set of active and rest states. The comparison of two SPECT examinations acquired respectively between (inter-ictal) or just after a seizure (ictal: radio pharmaceuticals being injected at the very beginning of a seizure), that respectively measures the inter-ictal and ictal regional cerebral blood flow, highlights areas involved in the epileptogenic network [6]. Finally, subtraction imaging is based on computation of differences between two data sets corresponding to two different examinations performed with and without tracer or contrast agent. In vascular imaging, this subtraction process extracts the blood vessels from the data sets. Intra patient and intra modality registration is also used to correct any patient movement during acquisition, especially for fMRI time series where a slight displacement during acquisition has a serious effect on statistical analysis [7-9].

Fusion Context II: Intra patient and inter modality registration

This is the registration between multimodal data sets concerning a single patient. Because no single medical imaging modality can detect or measure all anatomical structures or provide both functional and anatomical information with the highest spatial and temporal resolution, this fusion context is particularly useful in that it draws on the complementary nature of the different imaging modalities. It can be applied to many different anatomical areas (e.g. head, heart, breast [10], thorax (figure 11.2) [11], kidney, liver or entire abdomen [12], spine [13] and vertebrae, pelvis, limbs, eye fundus [14]).

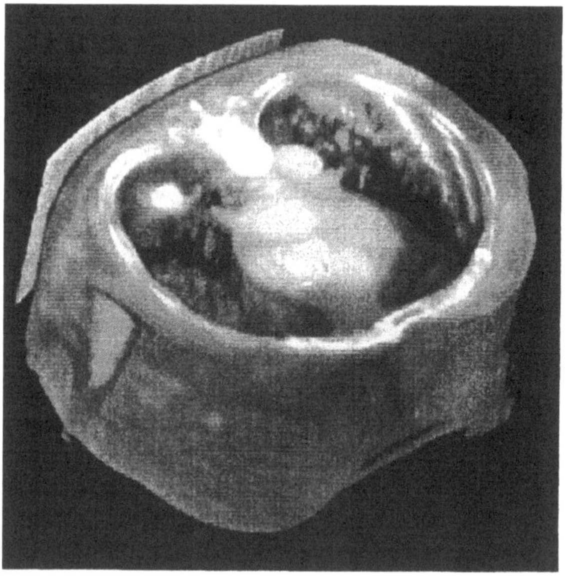

Fig. 11.2 CT/PET thoracic fused image of a patient with a lung cancer (courtesy C. Meyer, Digital Image Processing Lab., Dpt. of Radiology, University of Michigan Medical School, USA, [11])

For diagnosis, CT and MRI matching facilitates the anatomical localisation of both soft and bone tissues (figure 11.3) [15], and matching between fMRI, MEG, PET or SPECT and MRI allows anatomo-functional correlation (figures 11.4-11.6). Epilepsy surgery is a good example of multimodal surgery planning. Better understanding of the epileptogenic network and identification of the areas involved during seizures are possible from an analysis of multimodal information: anatomical (MRI) and functional (e.g. ictal and inter-ictal SPECT, figure 11.4), PET imaging, electrophysiological information (e.g. EEG, MEG, depth electrodes: SEEG or cortical measurements ECoG) as well as clinical investigations (e.g. interviews, videos, video-EEG) [16]. Another application of this fusion context is to use results from a data fusion process to facilitate image generation, reconstruction or analysis of a data set. An interesting example concerns the localisation of MEG Equivalent Current Dipoles from external magnetic field measurements (figures 11.5-11.7). This problem, known as an inverse problem, is not uniquely determined. Registration with other modalities creates additional constraints such as constraining candidate dipoles to be located within the brain grey matter (segmented from anatomical MRI), or to be located in brain areas identified as activated zones in fMRI using similar stimulation protocols [17]. During the performance of a neurosurgical procedure, soft tissues are subject to deformation (e.g. brain shift, tissue removal). Therefore, pre-operative images no longer correspond to the current patient anatomy. Intra-operative data may provide updated information about the location and modification of anatomy which, once registered with pre-operative images, may be used to compute deformations to be applied to pre-operative data [18-19].

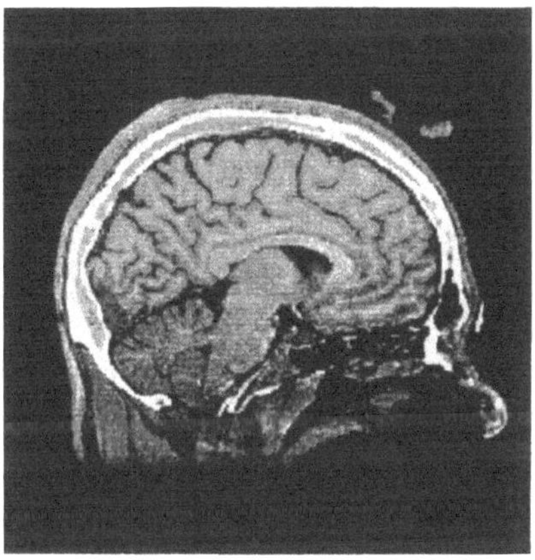

Fig. 11.3 MR/CT composite sagittal slice computed from a merged volume built by replacing into the MR data set voxels selecting from CT and belonging to the skull bone (courtesy J. Serrat, Centre de Visio per Computador, Universitat Autonoma de Barcelona, Spain [15])

Fusion Context III: Physical patient and modality registration

It is the registration between an image data set and the real physical patient. This fusion context addresses matching between the real patient and therapeutic tools (e.g. surgical tools, robots, interventional MRI, 3-D echography) and images to guide a surgical operation. The main clinical objective of this registration is to localise these images or measures in relation to the 'real' world (figure 11.8). 3-D localisers (e.g. mechanical [20], magnetic [21], and optical [22]) allow computing the patient location in the real world. Knowing this location (e.g. operating room, radiotherapy room [23] or imaging device room), one can compute the geometrical transformation between the patient position and the images. Neuro-navigation systems are the most commonly used image guided surgery systems (figure 11.9). This fusion context also concerns registration between images and tools. Stereotactic frames are used as instrument holders as well as markers and therefore have to be registered with pre-operative images in order to allow precise definition and performance of surgical trajectories for biopsies [24] (figure 11.10) or depth electrode implantation (SEEG) [25]. In neuro-navigation systems, the localizer may be able to track adapted surgical tools, with led or special patterns fixed on them, and to locate tip location and orientation in relation to images. Robotics applications may also require the same fusion context to achieve interaction between robots and images and/or robots and patients. Such robotics or neuro-navigation applications are mostly found in neurosurgery and orthopedics [26], but also in interventional radiology (e.g. endoscopy) or maxillofacial surgery. Interventional radiology may also consist of performing procedures under X-ray guidance (e.g. percutaneous laser discectomy, operations on arterio-venous malformations) [27-29]. The growing clinical use of image-guided therapy has enlightened clinicians on the additional benefits of using data fusion techniques.

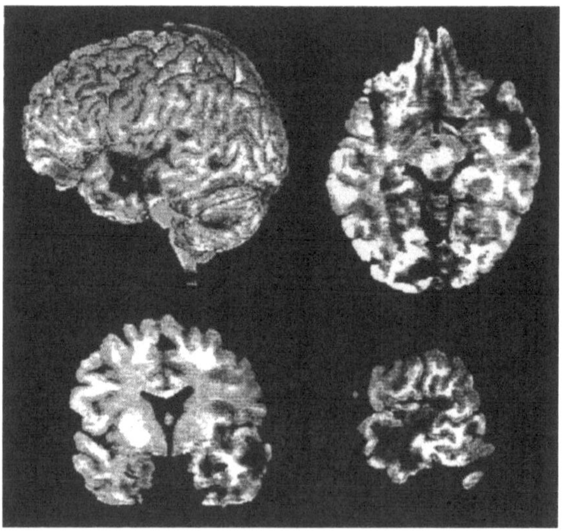

Fig. 11.4 MR/SPECT brain 3-D image and slices showing an hyper-intensity area related to the epileptogenic area detected by an ictal SPECT examination encoding by a 3-D coloured texture on MRI volume

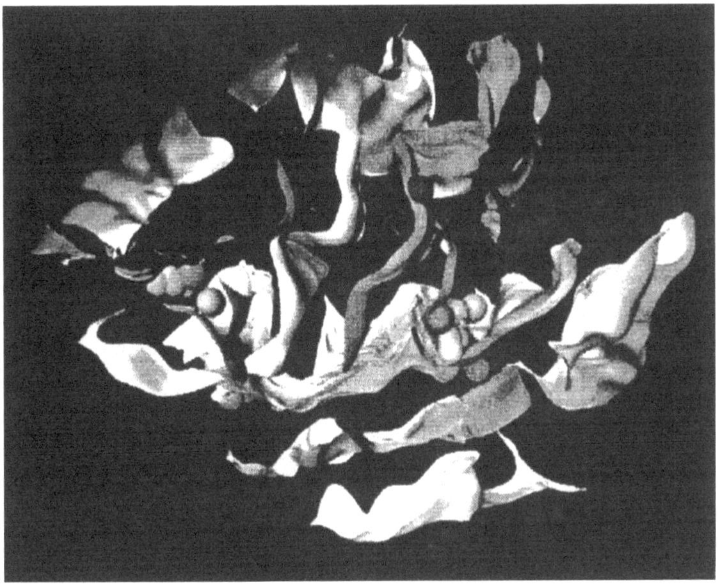

Fig. 11.5 Main cortical sulci of the left hemisphere of the brain segmented from a MRI data set and coloured spheres (colours encoding dipole latencies) representing MEG equivalent current dipoles corresponding of a vocalisation protocol

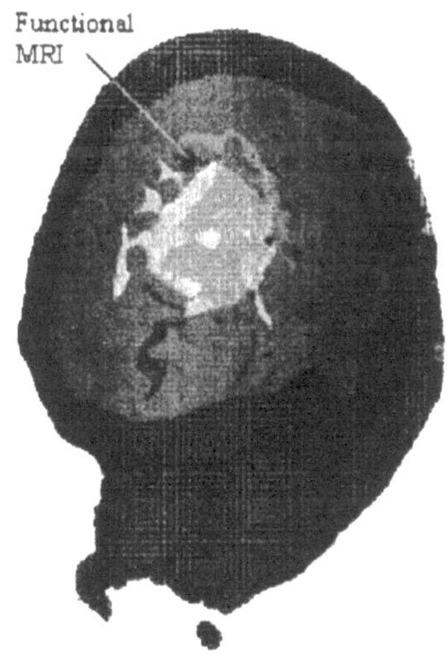

Fig. 11.6 3-D virtual scene of a patient including MRI information (skin, brain and tumour surfaces, and cortical sulci), MEG (black spheres) and functional MRI both representing language, motor and somato-sensory areas close to the tumour

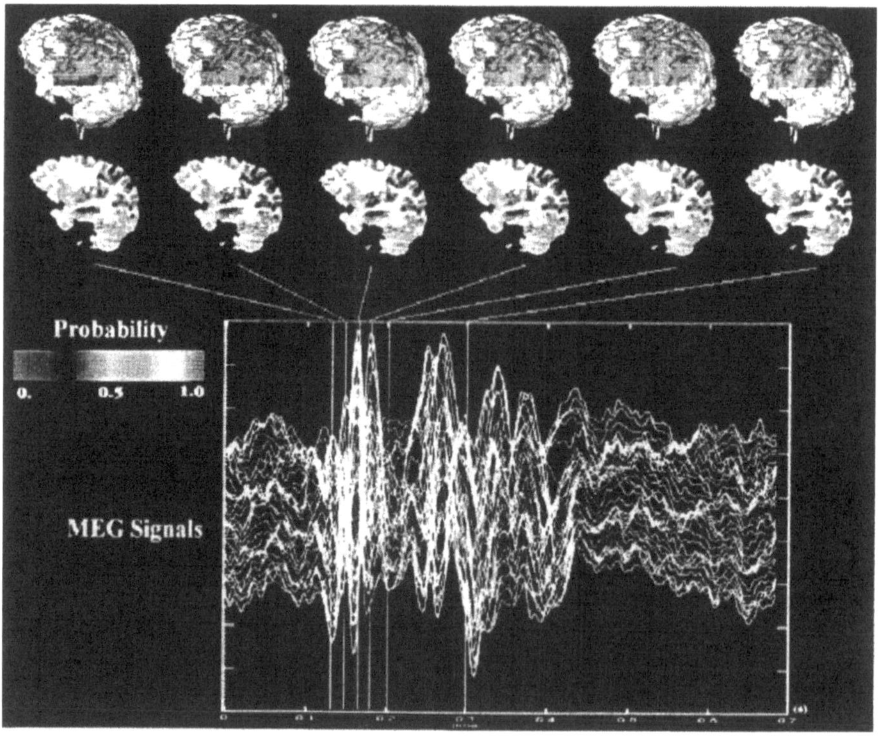

Fig. 11.7 MEG signals and the corresponding results of a spatio-temporal analysis of epileptic activity defining a region (top row) *posterior* to a cyst (middle row) displayed as a 3-D colour texture on MRI images

Fusion Context IV: Inter patient and intra or inter modality registration

This is the registration between a data set and multimodal data sets concerning different patients. The application of this fusion context is related to the study of inter patient anatomical or functional variability. It is used to compute average data (templates) or to compare patient data with an atlas or a template (figure 11.9). The construction of digital atlases requires registration of different patients in a common co-ordinate system (i.e. spatial normalisation) [30]. This fusion context also provides tools for model guided segmentation [31] or even model guided registration [32]. Registering different data sets of one modality from different patients may also improve the statistical significance of findings. Templates are used for statistical studies on morphology or pathology and comparison with normal subjects. Other applications are related to the functional mapping of the human brain gathering functional information from different functional modalities, and to anatomo-functional normalisation [33]. This fusion context is primarily used in brain studies although this approach could be extended to other anatomical structures. Methods and applications related to this fusion context are still being researched.

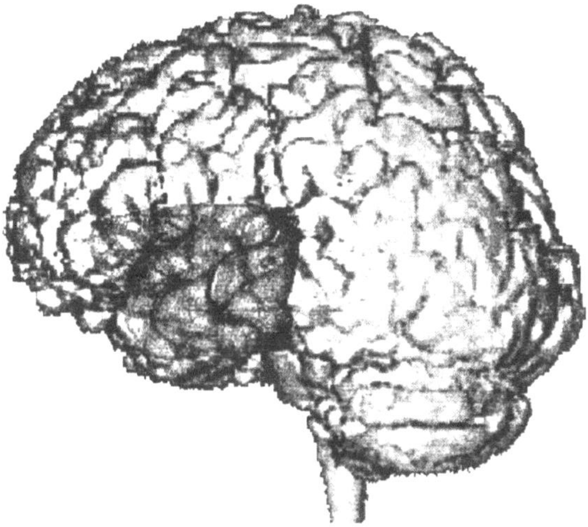

Fig. 11.8 Result of the registration of an intra operative view of the operating field with 3-D MRI

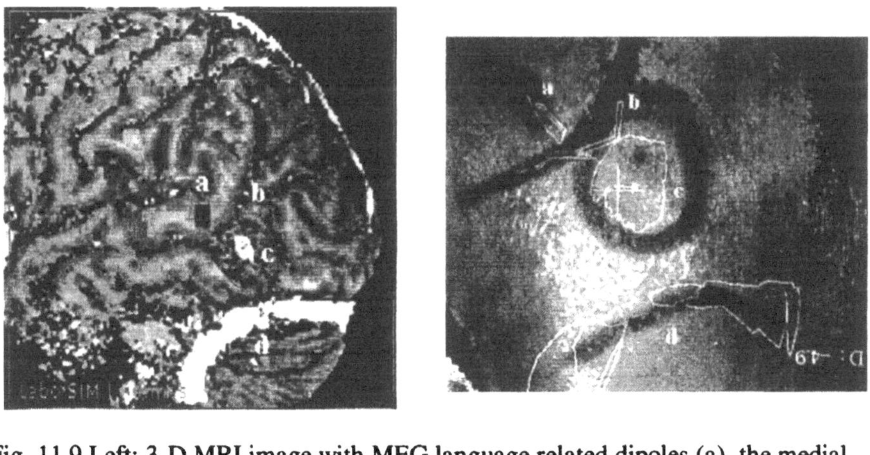

Fig. 11.9 Left: 3-D MRI image with MEG language related dipoles (a), the medial temporal sulcus (b), the meningioma (c) and the lateral sinus (d). Right: Intra-operative view through the oculars of a microscope with superimposed graphical information segmented from MEG (a), MRI (b, c) and MRA (d) data sets

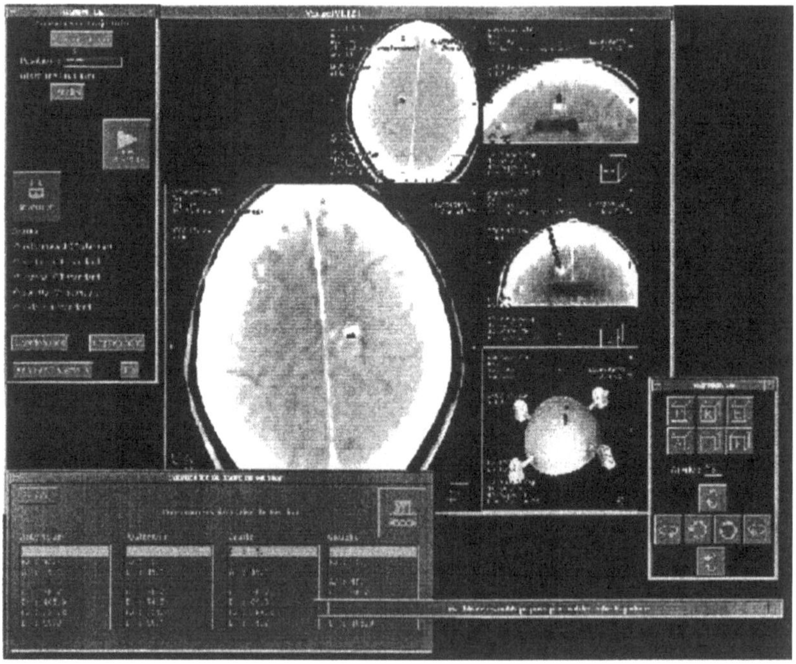

Fig. 11.10 Computer software for computer assisted stereotactic biopsies using the Fisher stereotactic frame and CT scan images

11.2.4 Data Fusion Methods

Data fusion methods are defined by paradigm and computation methods. The paradigm includes the fusion context and the clinical context. Both are intrinsic to the problem and they are the basis for choosing computation methods adequate for data fusion. Computation methods include registration, visualisation, interaction and analysis according to the results of registration. In the same way and at a lower level, registration methods are also defined by paradigm and computation methods. The paradigm includes a homologous information, which is assumed to be found in both data sets, and a model of the geometrical transformation best adapted to geometrical differences between data sets to be registered. Both are defined according to the paradigm of the data fusion method. The computation method is based on the paradigm of the registration methods and attempts to optimise the cost function, i.e. a similarity metric, measured on the homologous information via the transformation. In some situations geometrical transformations can easily be computed based on assumptions or constraints (e.g. no patient movement between acquisitions, well-known acquisition system geometry, robots or surgical tools related to imaging devices). In such situations the registration problem may have an analytical solution. In other cases, geometrical transformations have to be computed from fewer assumptions, therefore leading to an approximation of the real geometrical variations; however this approximation may be sufficient for data fusion applications. This aspect will be discussed in depth in section 11.5. Consequently, it appears that for a clinical application different data fusion methods may be used depending on the assumptions made and on the required approximation level.

In the next two sections, two basic components of data fusion computation methods, registration, matching methods, are described in more detail (figure 11.11).

11.3 REGISTRATION

Registration consists of computing a transformation between two data sets. This process compensates for spatial and intensity variations (e.g. due to different imaging devices and changes in acquisition geometry or conditions). Here we primarily focus on registration methods that account for spatial variations. Computation of intensity variation related transformations is mostly found in the fusion context *IV* [34-35].

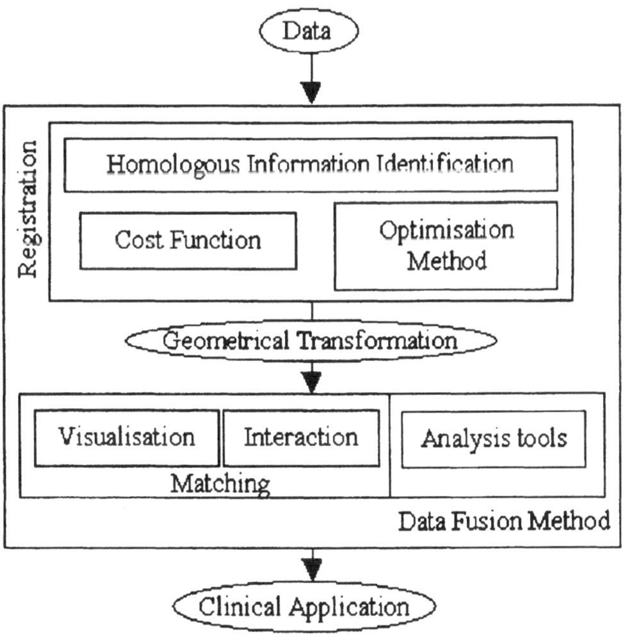

Fig. 11.11 Schematic of the data fusion procedure

Generally, registration methods require definition of a search space, a feature space, a similarity metric and a search strategy, as introduced by Brown [5]. Search and feature spaces refer to the registration paradigm as defined above. The class of transformations that models the variations between the data defines the search space and the homologous information between the data to be registered defines the feature space. The similarity metric and the search strategy refer to computation of methods. The similarity metric, or cost function, provides a measure of the quality of the matching for a given geometric transformation and a given set of homologous structures. The search strategy is consists of optimisation methods computing the best match according to this metric. In other words, the registration of two data sets is obtained by optimisation of a criterion of similarity, measured on homologous information extracted from the data. First, we find the geometrical transformation, which matches homologous information, and next it is assumed it applies to the data sets in their entirety. With each point of the first data set it associates a corresponding one in the second data set, in the common overlapping space. However this transformation is not necessarily bijective.

In most previous classifications of registration methods [1,3-4], there is not always a clear distinction between homologous information and cost functions. In our formalism, homologous information is a characteristic of the data in contrast to the cost function, which is not.

11.3.1 Data Co-ordinate Systems

In registration processes, definition of the different co-ordinate systems is crucial. Traditionally, in image analysis, three different co-ordinate systems or systems of reference are defined: the data co-ordinate system, the patient co-ordinate system and the world co-ordinate system. The data co-ordinate system refers to the pixel or voxel grid. The patient co-ordinate system is usually centred according to the data set and uses a standard length metric, deduced from the pixel or voxel size generally given by the imaging device. Finally, the world co-ordinate system is the combination of the patient co-ordinate system and information concerning patient orientation during acquisition. This information is also usually given by the imaging device and allows the registration procedure to be initialised. This world co-ordinate system is also fundamental to avoid left/right flip of patient data. Other application co-ordinate systems may also be defined (e.g. stereotactic frame co-ordinate system, Talairach atlas co-ordinate system) [34].

11.3.2 Registration paradigms

Homologous Information

Homologous information provides primitives on which search for the optimal geometrical transformation (i.e. the best match of this information) is based, according to a similarity metric. According to its structure, homologous information has to be defined (i.e. computed or segmented from each data set separately). Homologous information may be classified by its dimensionality: 0-D (point), 1-D (contour), 2-D (surface), 3-D (volume), or *n*-D (hypersurface). And by the dimension of its evolution space: 2-D (e.g. image, surface, projection), 3-D (volume, hypersurface), or *n*-D (hypersurface) with or without an additional temporal dimension [36]. Homologous information is frequently classified by its type, particularly extrinsic versus intrinsic information. Extrinsic information based methods use artificial and foreign objects attached to the patient and designed to be visible and easily identifiable in the data sets. Consequently, these methods require a specific image acquisition procedure with a dedicated protocol. Invasive techniques use bone screw mounted markers, stereotactic frames, ultrasonic or optical dynamic reference frames. Non-invasive techniques use dental adapters, moulds, frames, and skin fiducials. Extrinsic markers are specifically designed to be accurately detectable in both data sets to be registered. In practice, they can be extracted by automatic or semi-automatic procedures or by manual selection by a clinician.

Intrinsic information based methods use intrinsic patient information included in the data sets. Intrinsic homologous information may be composed of anatomical landmarks, geometrical features or information based on entire image voxel or pixel intensities. Anatomical landmarks refer to well-known anatomical points (e.g. anterior commissure and *posterior* commissure in brain MRI studies [37]), well-known anatomical lines (e.g. anatomical entity edges in 2-D images), well-known anatomical surfaces (e.g. skin, cortical sulci [38]) or to well-known anatomical volumes (e.g. brain, bones, vessels). Anatomical points are generally extracted manually, whereas anatomical lines, surfaces or volumes require segmentation procedures. Geometrical features refer to points, lines, surfaces or volumes that do not directly correspond to well-known anatomical structures (e.g. gradients, curvatures, local curvature extrema [39], vertices, crest lines [40]). Although considered to be homologous information, it is noted that geometric features can be used as an intermediate result for anatomical landmark segmentation (e.g. curvature measures used to segment cortical sulci). Anatomical landmark based homologous information is generally preferred to geometrical feature based homologous information as it always refers to *a priori* anatomical knowledge. Intrinsic homologous information based on entire image voxel or pixel intensities are preferred when common structures cannot easily be extracted from both data sets. The computation of this homologous information may refer to entire image voxel or pixel intensities (e.g. probability distribution, statistical measurements) or to intensities weighted by voxel location (e.g. principal axis, inertia moments). In the case of mutual information based methods, we consider homologous information as the distribution of the probability of image intensities, assessed by its histogram, and mutual information as a cost function that quantifies the statistical dependence between these distributions.

Geometrical Transformations

Geometrical transformations model spatial variations between the data. Geometrical transformations can be characterised by their nature (rigid, affine or non-linear) and by their application domain (local or global). In this sub-section, we define the spaces corresponding to the data to register as E1 and E2, and the geometrical transformation from E1 to E2 as T.

Linear Transformations

A T transformation is a linear transformation if and only if:

$$\forall x_1 \in E1, \forall x_2 \in E2, \forall c \in \Re$$
$$T(x_1 + x_2) = T(x_1) + T(x_2)$$
$$T(c, x_1) = c.T(x_1)$$

Linear transformations are used when *a priori* knowledge concerning registration problem geometry (e.g. acquisition conditions, rigid body properties of image entities) is available and justifies linear spatial variation assumptions. They greatly simplify the registration method.

A transformation T is affine if and only if $T(x)$-$T(0)$ is linear. The common form of an affine transformation between two N-dimensional spaces $E1$ and $E2$ is:

$$T(x) = Ax + b$$

where x and b are two N-dimensional column vectors, and A is an $N \times N$ matrix. An affine transformation can account for the most common spatial distortions as it is composed of a combination of N translations (vector b), N rotations, N scaling factors and N shear factors. In 3-D/3-D registration, affine transformations therefore need to assess twelve parameters.

In some medical applications, strong assumptions concerning data are sometimes made. For example, anatomical entities may be considered as rigid bodies, spatial sampling of the data may be known precisely and image distortions from imaging devices may have been corrected through a calibration process. Under these specific assumptions, a new category of transformations is commonly introduced as a rigid transformation. The rigid transformation is a subset of the affine transformation, composed of translations, rotations and an isotropic scaling factor. For 3-D/3-D registration this transformation can be assessed by seven parameters. Moreover, the scaling factor is often deduced from the spatial sampling of the data (e.g. voxel size), which brings the number of parameters to be estimated to six.

A global linear transformation is uniformly applied to the entire data set whereas a local linear transformation is defined on a specific subset of the data set. For instance, the Talairach co-ordinate system defines a piece-wise affine transformation consisting of twelve local linear transformations defined on twelve anatomically defined bounding boxes. This transformation is commonly used to register patients' images to the Talairach anatomical atlas [37].

Non-linear Transformations

Non-linear transformations model complex spatial and/or intensity variations. Among these, projective transformations are a specific case used when data is projected through an idealised image acquisition system [5,41]. One or more dimensions are lost via the transformation. This situation occurs essentially in the case of 3-D/2-D registration problems. Most non-linear transformations are referred to as elastic or curved transformations or deformations. Many functions can be used to model them, either by a single non-linear function defined over the entire data set (i.e. parametric functions), or by locally defined non-linear deformations, better suited to assess more complex variations. Geometrically the latter may be computed by using dense deformation fields δ where a local vector displacement is associated with each element of a grid defined on the data set; *E*1 and *E*2 are the homologous structures (i.e. homologous information) from both data sets:

$$\forall x_1 \in E1, x_2 = x_1 + \delta(x_1) \in E2$$

The definition of these deformations general includes a data related term (e.g. by using local similarity measurements as described in the next section) and a regularisation term to ensure spatial regularity of this transformation [36]. For 3-D transformation the computation of this deformation field may first consist of associating a grid with the volume using different decomposition schemes (e.g. finite difference model, finite element model). A displacement vector is associated with each node of the grid. Then, the regularisation produces a dense deformation field, which provides the required level of regularity (e.g. smoothing, continuity). Deformation models are globally non-linear but in practice can be locally linearised. Both the data related term and the regularisation term might be used to classify the different methods to model a deformation. Notably, we can distinguish landmark based approaches and iconic or intensity based approaches. Landmark based deformation models consist of using features extracted from both data sets (e.g. anatomical points, sulci or ventricle surfaces) to compute the deformation field. Interpolation methods are used to calculate the deformation over the entire volume while achieving regularisation of the deformation field. Among these methods, landmarks may consist of points (regularisation being achieved through interpolation based on radial basis functions such as the thin plate spline [42-43]), crests line [40], surfaces (regularisation obtained using mechanical elastic models [44], super-quadrics [45] or spline deformations [46]) or variation modes of an entity inferred from a statistical study. Among deformation models based on iconic approaches one may find multiscale deformations (e.g. ANIMAL [47], optical flow measurements [48]), the 'daemons' method [49-50], expansions on smooth basis functions (e.g. Fourier modes) that can model spatial and/or intensity variations [34] or models using local similarity measurements assessed using image intensities with regularisation models based on mechanical properties [51-52] (e.g. elasticity, plasticity, viscosity, fracture, fusion), or fluid properties [53-54].

One drawback of almost all the previous models is that they are based on normal anatomy and physiology (i.e. non-pathological) and so registration methods based on these models may fail when abnormal structures (e.g. tumours in MRI or CT, epileptogenic areas in ictal SPECT) are encountered.

Non-linear transformations are mostly used in two types of situation. First, when the entities to be registered do not refer to the same anatomical substrate (e.g. inter-patient image registration, image to atlas registration, image to model registration). Next, when a common anatomical substrate exists but might be disrupted by some non-rigid deformation (e.g. correction of acquisition deformation, intra-operative deformation correction, registration of several data sets showing a deformation or movement of the structure: growth, pathology, deformable anatomical structure (e.g. heart, lung, movements due to respiration).

11.3.3 Registration Computation Methods

Cost Function or Similarity Measurement

The cost function is an objective criterion used to estimate the quality of the registration. It is defined according to homologous information and to the model of the transformation, i.e. according to feature space and search space. Using Brown's terminology [5], the cost function defines a similarity metric that measures the quality of the transformation estimation, given a search space and a feature space. Two main categories of similarity metrics may be distinguished:

those based on Euclidean distance measurements and those based on statistical similarity measurements (table 11.1).

Using extrinsic homologous information, the definition of a cost function may depend on the geometry of extrinsic markers, particularly when using stereotactic frames. Nevertheless, this definition often leads to a mean Euclidean distance estimation. Point fitting problems (fiducial markers, landmarks) are generally solved by least square methods. For more complex segmented structures, specific algorithms have been developed using the Euclidean distance criterion to match complex surfaces; e.g. distance transform as chamfer distance matching techniques [55-56] or 'head-hat' method introduced by Pelizzari *et al.* [57]. The Euclidean distance between points is computed in a continuous space, whereas the chamfer distance is computed on a discrete grid in a discrete space. Such methods have proved efficient and robust for multimodal registration using skinhead surfaces segmented from MRI, CT or PET data. The skin surface is easily identifiable in such modalities and sufficient to constrain the assessment of a rigid transformation. When more complex models of transformation are linearised, (e.g. expansion on smooth basis functions), the cost function only consists of a least squares formulation [34].

Table 10.1 Summary of usual cost functions

Similarity type	Similarity metric	Similarity properties
Euclidean distance	Least squares distance Chamfer distance transform	Continuous distance measurements Discrete distance measurements
Image voxel or pixel intensities based measurements	Cross-correlation coefficient Correlation ratio Woods criterion Mutual information	Linear dependence Functional dependence Uniformity properties No assumptions on the nature of statistical dependence

When comparing homologous information based on intensities distributed across the entire image, Euclidean distance criteria are no longer suitable, and criteria based on statistical similarity measurements are generally preferred. Similarity criteria generally make some assumptions on the statistical dependence of the data. For example, the cross-correlation coefficient measures linear dependence between both data intensity distributions, whereas the correlation ratio measures functional dependence [58-59]. The Woods criterion [60] is based on the assumption that a uniform intensity region in one data set matches another uniform intensity region in the second, and finally mutual information [61-62] measures the statistical dependence of both data sets without any assumptions on the nature of this dependence. Thus, *a priori* knowledge concerning data to register has to be taken into account when choosing a statistical similarity measurement [63]. For example, the cross-correlation coefficient may be used for monomodal registration (realignment of time series in fMRI). Mutual information is effective for both monomodal and multimodal registration especially when little knowledge concerning statistical dependence is available. Woods' assumption on uniformity is well adapted for MRI/PET registration [60]. Many other similarity measurements exist, each one imposing specific assumptions on the nature of the dependence: variance ratio, histogram analysis, zero crossings in difference images, absolute or squared intensity differences, optical flow, entropy. Such measurements may be computed either in the spatial domain or in the Fourier frequency domain. For example, a simple transformation such as a single translation can be directly assessed using the Fourier transform properties (e.g. phase correlation techniques [64]).

Methods based on similarity measurements operate on image intensities and consequently are sensitive to interpolation methods used to compute the corresponding signal values in the target data set [65], particularly because signal values are only available on discrete grids whereas transformations and also transformed co-ordinates are computed in a continuous space. Most of the similarity criteria listed above are computed via joint histograms, whose computation requires interpolation methods. Different interpolation methods may be used such as, for example, nearest neighbour, tri-linear, or partial volume interpolation, with different characteristics. Nearest neighbour interpolation does not allow sub-voxel registration accuracy, whereas tri-linear interpolation does but creates artificial grey level values during joint histogram computation and thus perturbs the intensity probability distribution assessment. Therefore, partial volume interpolation is generally preferred. Actually, this technique does not introduce artificial grey levels since increments in the joint histogram are the weights computed in the tri-linear interpolation. In the latter case, the joint histogram and the corresponding similarity measurements (e.g. mutual information, correlation ratio) become continuous functions of the transformation parameters and their partial derivatives can be computed [66].

Another approach is to combine different cost functions or higher order information from *a priori* knowledge to improve the registration procedure. For example, *a priori* models of brain tissues may be used to determine the inter-subject elastic registration [34]. Sometimes similarity measurements must be able to cope with data intrinsic dissimilarities. For example, ictal SPECT/MRI registration based on statistical similarity measurements may be disturbed by SPECT hyper intensities related to epileptic seizure (e.g. high regional Cerebral Blood Flow during seizure), since these hyper intensities have no equivalent in MRI. Some techniques allow cost functions to account for such dissimilarities, in order to reject them from similarity measurements automatically: outlier rejection technique such as outlier processes or robust estimators [67]. In the context of ictal SPECT/MRI registration, Nikou *et al.* [68] propose a solution making use of robust estimators on statistical similarity measurements (least squares norm or Woods criterion).

Optimisation Methods

The search strategy defines the method to achieve optimal matching between both data sets, i.e. to compute the best transformation according to a similarity metric measured on homologous information. Accordingly to available assumptions, the optimisation problem may have an analytical solution (e.g. least squares solution) or require iterative optimisation methods. Many of these optimisation methods can be found in Press *et al.* [69]. The choice of an optimisation method is an algorithmic choice that depends on the cost function and its mathematical properties rather than the clinical context. Below is a classification of optimisation methods in the field of medical data registration problems as introduced by Barillot [36]. Four approaches can be distinguished:

- Quadratic or semi-quadratic approaches: these methods are based on the assumption of convexity (e.g. gradient descent, Powell, Newton-Raphson, Simplex) or quasi-convexity (Iterative Closest Point: ICP) of the cost function near the optimal solution. These methods are well adapted for well-constrained problems (e.g. unity of the solution). Cost function properties, particularly differentiability and regularity, generally determine the choice of optimisation method.
- Stochastic or statistical approaches: these methods (e.g. simulated annealing, Iterative Conditional Modes: ICM, genetic algorithms) are particularly effective in dealing with outlier data (e.g. noise) during optimisation and are therefore more reliable for solving under-constrained problems and avoiding local optima. Nevertheless, improvement in terms of robustness generally implies low performance in terms of computation time.
- Structural approaches: these methods rely on tree or graphs based optimisation techniques (e.g. dynamic programming), and require an exhaustive study of the entire search space to find the optimum. Thus, a global optimum is always reached, but this generally requires significant computation time. Actually, structural methods are well suited when similarity measures can be formalised by a hierarchical structure.
- Heuristic approaches: such methods generally consist of interactively finding solutions by visually optimising similarity measurements.

Moreover, defining an optimisation strategy may solve common optimisation problems, such as avoiding local extrema or reducing computation time. Multi-resolution or multiscale strategies better match each level of analysis to the constraints and assumptions required by the optimisation algorithm. The optimisation strategy may also include suitable initialisation of the search, adaptation of constraints during the optimisation process (e.g. refinement of the results) or use of a combination of several optimisation methods. For example, in the context of non-linear registration most methods initialise the registration process by computing a global affine transformation.

11.4 MATCHING AND ANALYSIS: USING THE RESULTS OF REGISTRATION

Once the data are registered, i.e. the transformation is computed, it is possible to compute for each point of one data set, the co-ordinates of the corresponding point in the other data set. However, tools are required to further exploit results of the registration. Generic visualisation and interaction tools are essential in all clinical applications of data fusion to allow matching and qualitative interpretation and exploitation of the registration results. More specific analysis tools may also be required to enable a more accurate and in-depth exploitation of the results or to support a more quantitative interpretation.

11.4.1 Visualisation and Interaction

As said previously, clinicians did not wait for the development of computerised data fusion tools to perform multimodal matching. Indeed, they actually matched data mentally. The techniques of data fusion can directly provide visual access to the data fusion results. One use of these visualisation tools concerns the checking of the registration solution. These tools may implement either of the following two approaches: a 'visual superimposition approach' or a '3-D-cursor approach'. Both may be described according to the information shared between the data sets (e.g. a point, a contour, an image, a surface or an entire volume), to the graphical representation used to visualise this information (e.g. graphics, volume, texture, transparency, colour encoding) and the display devices (e.g. computer screen, head mounted display, microscope) [70-71].

Visual Superimposition Based Approach

The aim of this approach is to compute and display a single image featuring information from registered data sets. Several methods can be used. In one solution, multimodal data sets are superimposed pixel by pixel into a single image (figure 11.4). Superimposition can be done by arithmetic operation on pixel values (e.g. transparency), colour encoding, or spectral encoding [72-74]. For instance, for MRI/SPECT data fusion, an high-resolution functional image may be computed using fuzzy brain tissue classification and fuzzy fusion operators [75]. When more than two data sets are displayed together, this abundance of information may impair image readability. Techniques such as 'magic window' (i.e. area of interest defined and moved by an operator in which the original pixel values of one data set are replaced by those of the other one) or 'chessboard' (i.e. alternating pixels) are useful for examining data in 2-D, but difficult to apply to 3-D display. Based on a similar concept, a merged volume may be built by selecting the voxels corresponding to relevant structures from each data set (figure 11.3). These two visualisation modes (i.e. superimposition or replacement) may be computed in 2-D or in 3-D using volume-rendering techniques. One drawback of these methods is that the volume computed from a combination of data sets no longer corresponds to a real modality and this may confuse a clinician. Another solution for reducing the complexity of the scene is to extract surfaces from most relevant structures from different data sets and represent them in a new 3-D scene using surface rendering techniques. For instance, if the clinical objective is to study the functional activity of specific cortical areas, we may use the surface of the sulci segmented from MRI to represent the frontiers of these anatomical areas, and the surface of spheres depicting the locations of equivalent current dipoles from magneto-encephalography to represent the functional areas (figure 11.5).

This visual superimposition based approach requires the computation of new data sets resulting from the transformation of original data sets via the geometrical transformation computed by the registration method. Volumes have to be reformatted and therefore interpolation methods may have an impact on the display (e.g. discontinuities) and should be chosen carefully.

Cursor Based Approach

When fusion is required between more than two data sets, the superimposition approach may not be the most appropriate solution. The cursor-based approach allows matching between several data sets merely by sharing a cursor between synchronised display modules. A mouse event in an image corresponds to the designation of a point in the corresponding data set. The co-ordinates of this point are sent to the other display modules and transformed according to each geometrical transformation, which triggers the display of the images containing the corresponding point within each module [71,76].

The superimposition approach facilitates comprehension of multimodal information notably by achieving global realignment of both data sets on a common grid. The cursor approach makes it possible to display as many modules as required to examine the different data sets. With this approach, global realignment (i.e. data reformatting) is not required. However, if it is not performed, matching will be limited to the region surrounding the cursor. From our experience [75], it is clear that visual superimposition and cursor approaches are complementary and should be selected according to clinical objectives.

Analysis Tools

Visualisation and interaction tools allow a qualitative interpretation or analysis of the results of the data fusion process. Some applications require the clinician to go beyond this qualitative interpretation by performing quantitative processing on registered data sets. This additional category of tools may be considered as a new component of data fusion systems. In some applications, quantification aims at analysing the intensity related variations. For instance in epilepsy, the subtraction of ictal SPECT from inter-ictal SPECT enhances areas involved in epileptic phenomena. This subtraction process requires pre-normalisation of both data sets (SISCOM: Subtraction Ictal SPECT Co-registered with MRI [77-79]). Data fusion may also be an intermediate step in the course of an image analysis process, such as a segmentation process (e.g. atlas based segmentation, multi-spectral classification of brain tissues in MRI), or an image reconstruction process (e.g. computation of MEG equivalent current dipole location constrained within the brain grey matter segmented from anatomical MRI, same computation constrained within activation areas detected from fMRI studies [17]).

Statistical analysis may be used for a quantitative comparison of different data sets, e.g. in anatomical inter individual variability studies (after inter patient registration [80]) and in brain functional mapping studies [81]. The latter quantitative comparison may concern one or more subjects and one or more modalities and is particularly relevant in neuro-imaging.

Originally, data fusion was limited to registration, i.e. computation of the geometrical transformation. Additional visualisation and interaction tools were rapidly considered necessary for a visual or qualitative analysis of the matching. Recently analysis tools have added a quantitative aspect to the analysis of data fusion. From this development, one can reasonably imagine that data fusion tools will go on evolving towards decision support components offering enhanced capabilities (e.g. high level statistical analysis, knowledge based tools, connection with local or remote clinical case repositories) (figure 11.12) [82].

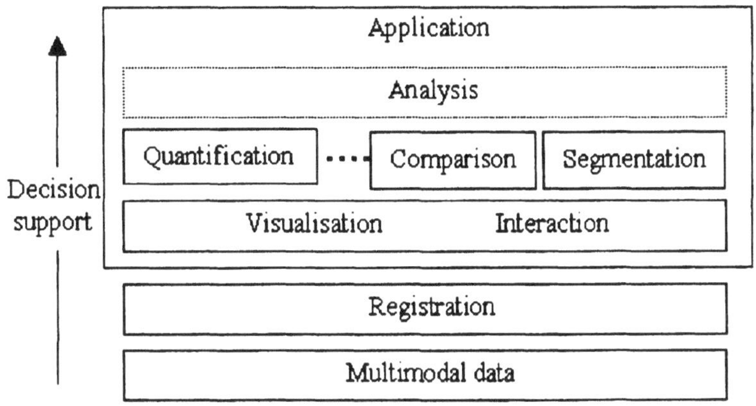

Fig. 11.12 Evolution of data fusion systems

11.5 VALIDATION

Validation is an essential step in the implementation of a data fusion method. Validation is required to describe and analyse the characteristics of a method, to demonstrate its performance and potential advantages and to compare it with existing methods. Validation has to be performed once and is a prerequisite for any clinical use. Besides, another aspect of validation has to be taken into account. It concerns the checking procedure during any daily clinical use of data fusion method.

Errors may occur at every step of a data fusion process (table 11.2), therefore validation must be performed for each component of the procedure and not be limited to registration, although registration obviously plays a prominent role in the quality of the data fusion process. A validation procedure generally addresses the following topics: precision, accuracy, robustness, complexity and computation time. However, these topics are not independent and may require some tradeoffs. The next sections introduce these topics and describe some validation procedures. These are principally illustrated by the case of registration validation but they may also be used in the other stages of data fusion processes (e.g. acquisition, reconstruction, segmentation, visualisation). The importance of checking procedures during any daily clinical use of data fusion is defined and emphasised.

Table 11.2 Potential sources of errors in data fusion methods

Sources of errors	Description
Image acquisition	Artefacts, image distortion [83], patient movement [8], image resolution, reconstruction process
Definition of the homologous structures	Human error in manual identification of markers or landmarks, segmentation errors
Residual errors	Estimation error during optimisation of the cost function
Geometrical transformation	Errors caused by interpolation, reconstruction, reformatting process
Round-off errors	During any computation
Misinterpretation	During visualisation and/or interaction with registered data [74]
Model unsuitability	Similarity assumptions no longer valid due to clinical modifications (e.g. evolution of a tumour, pre/post surgery, effects of treatment)

11.5.1 Precision and Accuracy

In this section precision is distinguished from accuracy as follows. The precision of a process is related to its resolution: the minimum value of the systematic error that we can expect from this process. Precision is intrinsic to this process. This value is generally expressed in the parameter space. The spatial precision of the image acquisition step corresponds to the spatial resolution of the images. The precision of the homologous information definition is related to the precision of the manual designation or of the segmentation procedure. The precision of registration is related to the minimum variation of the transformation that the system is able to detect. The precision of visualisation corresponds to a combination of the resolution of the displayed image, the eye resolution, and the computer screen resolution. By applying a specific variation to the input, one can verify the ability of the process to recover it.

For each voxel of the data set, the local accuracy is defined as the difference between the computed co-ordinates and the actual (i.e. theoretical) ones assumed to be known from a gold standard. This difference is generally referred as the local error. A global accuracy value can be computed for the entire data set from a combination of local accuracy values (e.g. mean and standard deviation). In some specific cases a spatial distribution of the accuracy within the entire data set can be computed from the local accuracy values under specific assumptions (e.g. continuity).

However, clinicians need to be aware of the precision and accuracy of the entire data fusion process, rather than the specific precision and accuracy values of each step. It provides them with an overall idea about the confidence limits, and of what may be expected from the system, according to clinical objectives. In practice, clinicians are not always informed about system precision. Some data fusion systems provide a separate precision value for each step (e.g. acquisition, reconstruction, registration, localisation) but seldom a global precision value for the entire process. Similarly, most data fusion applications provide an accuracy value, which only concerns the registration method. Defining a precision or accuracy value for an entire data fusion process is a difficult task and, to the best of our knowledge, no general formalisation of this problem has yet been proposed.

11.5.2 Robustness

A method should obviously be sensitive to slight variations, which need to be compensated for but not sensitive to disruptive factors (e.g. noise, presence of lesion). For instance, the robustness of a method can be assessed by checking its behaviour when simulated disruptive factors are injected into the data.

11.5.3 Complexity and Computation Time

Complexity and computation time are characteristics of method implementation. Algorithmic complexity can be determined analytically and has a direct impact on computation time. Functional complexity concerns the steps that are time-consuming or cumbersome for the operator. It deals both with man-computer interaction and integration in the clinical context. For example, assessing the functional complexity of a registration method consists of studying the following steps: pre-processing operations, thresholding, definition of the homologous structures, registration initialisation, and stop criterion.

The degree of automation of the method is an important aspect of functional complexity (manual, semi-automatic or automatic). Automatic methods only require designation of the data to be registered. Semi-automatic methods require user initialisation of the geometrical transformation, and/or a manual designation of the homologous structures and/or a manual check on results. Manual or interactive methods involve a totally manual process under visual inspection of the user. In most methods, there is a trade-off between degree of automation, speed, accuracy and robustness. For instance, some methods might benefit from user intervention - consisting of correctly initialising the process, or checking the search directions, or rejecting aberrant solutions at a low-resolution level - allowing better convergence of the optimisation algorithm.

One should never forget that a data fusion process is only one step in the course of a clinical procedure. The impact of complexity and computation time will be different according to the clinical context. For instance, off-line registration puts fewer constraints on the user whereas on-line registration requires fast and automated procedures.

11.5.4 Implementation of Validation

There are different methods to implement a validation procedure, depending on the features to be validated. Precision values of a system are estimated by analysing its behaviour under controlled experiment conditions with validation data sets (table 11.3). Additional validation methods may be used for computing accuracy values as well as studying robustness: multi-observer visual assessment, receiver operating characteristics (ROC) curve studies and comparison with reference software packages. Complexity and computation time may be assessed and compared with common reference software.

Table 11.3 Validation data sets

Computer simulated data sets
Phantoms (well known geometrical shapes)
Reference data sets (Vanderbilt, visible human project)
Realistic data sets (cadaver, clinical data sets)

Validation is obtained through comparison with a gold standard (e.g. difference between the computed location and its actual location assumed to be known from the gold standard). The gold standard is given by well-known locations of entities within the data (e.g. fiducials markers, frames, probed points, and anatomical landmarks) or by controlled experiment conditions (e.g. simulated data, phantom). Data obtained from numerical simulation theoretically allows all parameters to be checked (geometrical transformation, noise level) but may be far from the clinical reality. In this context, validation based on data simulated from patients' high-resolution data sets might be a well-suited solution. Any protocol and/or modality could be simulated to produce realistic data sets, which can be further altered to fit imperfect settings. Phantom-based data account for acquisition conditions but are far from the metabolic or physiological reality [84].

Reference data sets allow validation in more realistic conditions, as demonstrated by the Retrospective Registration Evaluation Project (RREP) conducted by Fitzpatrick *et al.* at Vanderbilt University [85-86]. This project consisted of a blind evaluation comparing registration techniques from different research groups. All validations were performed on the same data sets, in reference to registration gold standards established using a stereotactic frame. Validation on data sets from living patients is close to the clinical reality but patient movement problems cannot be avoided even by using fiducials markers.

Movement problems can be avoided by using cadavers instead of living patients, but this data cannot be used when functional or metabolic information is required. Validation data range from very realistic data (i.e. close to the clinical reality), but still very sensitive to experiment conditions to artificial data (e.g. numerical simulation), which is easier to control. Visual assessment by clinical or non-clinical observers provides a qualitative approach for validation. Its principle is to ask operators to visually and repeatedly assess the quality of a procedure result (e.g. registration, segmentation). This visual assessment can be performed using interaction and visualisation tools as described in section 11.4.1. Receiver operating characteristics studies [87] may be used to quantify this operator based assessment by comparing it with a ground truth.

A common practice within the data fusion community is to compare new methods with well-established standard software (e.g. SPM, AFNI, and AIR, *cf.* Appendix). These software packages are generally well documented and result either from common research site specifications or quasi-industrial developments. Because they are widely used, they contain experiences from many different clinical contexts.

As a concrete illustration of validation implementation, we will describe the Maes *et al.* experience reported in [66]. The authors describe quantitative validation of a specific aspect of the registration step, namely the choice of an optimisation method and a search strategy. The context of this study is multimodal registration (CT/MRI and T1 MRI/T2 MRI) using mutual information computed with partial volume interpolation. Validation is based on data sets from the Vanderbilt database [85-86]. Different optimisation algorithms (e.g. Powell, simplex, downhill gradient, conjugate gradient, Newton-Raphson and Levenberg-Maquard) and different search strategies (different 2- or 3-level multiresolution schemes) are compared. Precision is assessed by the variance of the solution in the parameter space within a homogeneous population of tests. Accuracy is measured from misalignment of well-known points, whose actual locations on the brain surface are given by the gold standard. From these measures, a mean value gives an estimation of a global registration error. Algorithm performance in terms of computation time is evaluated using the number of cost function to reach convergence.

11.5.5 Checking Procedure During Clinical Use

Validation only characterises the performance of a data fusion method through controlled experiments. However, it cannot guarantee satisfactory behaviour of the data fusion method in each clinical use. Registration algorithms generally provide the user with residual errors resulting from the optimisation procedure. However, these values may not be reliable indicators of registration quality since the solution may be mathematically optimal, but not relevant from a clinical viewpoint (e.g. local optima). Therefore, clinicians should visually verify the registration results before any clinical use of the data fusion system. They should check and explicitly accept the results of each step of the data fusion procedure separately. However, it may not be easy for clinicians to define the criteria driving this decision.

For instance, the global accuracy of registration may be assessed by verifying the location of well-known points on both registered data sets, irrespective of the residual errors computed by the registration algorithm. For instance, in neuro-navigation systems, the surgeon can check registration accuracy by pointing out well known anatomical points on the patient in the operating room and verifying the corresponding locations on the pre-operative images. A classic mistake is to make this verification on the fiducials markers or the homologous structures used for the registration (homologous structures should never be used in the validation process). This checking procedure detects cases where a mathematically correct but geometrically aberrant solution is found (e.g. local optimum of the cost function). By visualising and analysing fused images the surgeon can also visually assess the quality of the registration. This checking procedure can be performed using interaction and visualisation tools as previously described in section 11.4.1.

Throughout the process, data fusion systems should also visually integrate precision and accuracy values, in order to explicitly remind users of the level of confidence to be expected in the fused data. For example, a fuzzy representation of the fusion or a confidence area may be displayed instead of a single location [88].

11.6 CASE STUDIES

In this chapter, a recently developed medical application of data fusion concerning magneto-encephalography (MEG) and anatomic MRI data fusion [89-90] is presented. This research work was performed in the SIM laboratory of the University of Rennes in France, in collaboration with the University Hospital of Rennes, 4-D Neuroimaging of San Diego in the USA, and ETIAM of Rennes.

The clinical context of this MEG/MRI data fusion is related to functional brain mapping and particularly anatomo-functional correlation studies. The anatomical area involved is the brain and this application refers to a specific clinical procedure performed on one patient. MRI data provides information about brain anatomy whereas MEG measures the neuronal magnetic activities related to specific cerebral functions (e.g. somato-sensory, motricity, language) or to specific cerebral phenomena such as epilepsy. Thus, the clinical objectives of this medical application of data fusion concern either pre-surgical functional brain mapping during the planning of a therapeutic procedure, or the characterisation of the epileptogenic network. MRI data consist of 3-D acquisitions using a T1 weighted spin echo sequence acquired on a Signa 1.5 T (General Electric Medical Systems).

MRI 2-D slices are then interpolated to compute a high-resolution isotropic 3-D volume (voxel size = 0.94 mm). The MEG examination consists of recording the magnetic fields outside the head using a 37-sensor MAGNES system (from 4-D Neuroimaging). It provides measurements of brain activity with a temporal resolution of few milliseconds. One can attempt to localise the sources of the recorded activities, and estimate spatial and temporal characteristics of such sources modelled by equivalent current dipoles. This inverse problem is solved using a spatio-temporal approach [91].

This MEG/MRI data fusion problem belongs to fusion context II, i.e. an intra-patient inter-modality data fusion problem. It involves MRI and MEG equivalent current dipoles, which is a 3-D/4-D data fusion context. The patient's head is considered as a rigid body; this is the only assumption of similarity we can put forward about MEG and MRI data.

For the registration step, three co-ordinate systems have been distinguished: a world co-ordinate system related to MRI data, a MEG sensor or acquisition co-ordinate system, and a MEG patient co-ordinate system defined by three anatomical landmarks: the nasion, the left pre-auricular point and the right pre-auricular point. These points are referenced in the MEG sensor co-ordinate system using a Polhemus 3-D magnetic localizer. Concerning the registration paradigm, homologous information is provided by the skin surface. This surface is extracted from the MRI 3-D data set (threshold, connected component analysis and morphological operators) on the one hand and measured in the MEG sensor reference system by acquiring a head-shape with the Polhemus 3-D digitizer on the other. A rigid transformation is chosen to model the geometrical variations. The registration method consists of a cost function based on the Euclidean distance between both surfaces that is minimised using Powell's optimisation algorithm and a multiresolution search strategy. A chamfer distance transform is computed from the skin surface extracted from the MRI data set, which gives an approximation of the Euclidean distance from the skin surface for each voxel of this data set [55]. The cost function is computed as the average of the chamfer values of all the MEG head-shape points after transformation into the MRI co-ordinate system. A robust estimation of the cost function is also achieved by automatically rejecting head-shape inconsistent points during optimisation.

Different visualisation and interaction tools are then used to verify registration and to perform anatomo-functional mapping. One visualisation mode consists of superimposing source locations as points or vectors (dipole location and orientation) on MRI data, allowing interaction with dipole characteristics (e.g. latency, signal RMS, localisation quality criteria). This fusion context also allows co-operative data fusion: after registration grey matter segmented from MRI data may be used to constrain MEG source localisation. Thus, at each signal latency a 3-D colour texture representing source existence probability at each grey matter point can be superimposed on MRI data (slices and 3-D rendering) [71]. Animation may also be generated to better appreciate the high temporal resolution of MEG (figure 11.7). Surface rendering is also used to get maximum benefit from the spatio-temporal information provided by MEG and MRI and to focus on the most relevant features (e.g. most significant dipoles, cortical sulci) (figure 11.5) [92].

A validation procedure was performed to demonstrate the theoretical and experimental performance of the method in a well-defined clinical context. Potential sources of error were studied, i.e. measured, modelled and simulated for each step of the process, from data acquisition to data registration. Concerning MRI data acquisition and head surface segmentation, errors may be classified as scaling effects (inherent to acquisition and surface segmentation), local artefacts (eye movements, dental prosthesis) and noise. As acquisition artefacts due to dental prosthesis are impossible to remove, image data located below the nose was not taken into account for registration. Similar types of errors may affect head-shape digitisation: scaling effect (skin mechanical properties), local artefacts (especially due to manipulation errors: patient movements, operator errors) and noise. The Polhemus digitizer's accuracy was precisely measured: with or without human manipulation (0.16 ± 0.07 mm *vs.* 0.20 ± 0.04 mm), on a phantom (0.80 ± 0.07 mm) and on a real head (1.50 ± 0.43 mm). The theoretical validation of the registration was implemented using computer simulated data sets: a theoretical head shape that perfectly matches an MRI data set was constructed, with additive noise reflecting different sources of error modelled from the previously described measurements. Global accuracy values were estimated by computing the mean real error, its standard deviation, its minimum and maximum values on the overall MRI data set (in a 15.00 cm side cube sampled every 2.00 mm). Different parameters were evaluated separately: the algorithm's ability to detect a translation, a rotation or a combination of both transformations, the effect of the number of points of the head shape, the effect of noise, the effect of inconsistent points and the effect of scaling. Finally, more realistic simulations were performed by combining all these parameters. The accuracy estimation showed excellent reproducibility of results with a low mean registration error of 3.00 mm. Clinical validation was performed by comparing this automatic algorithm with a clinically used reference technique: a manual registration technique. After transformation the head shape points were superimposed on the MRI data set. An expert evaluated the quality of the registration produced by each method at each point of the head-shape (0 = bad registration, 1 = good registration). A sign test performed on 13 patients showed that automatic surface matching based registration is significantly better than manual registration. The reader will find an exhaustive description of these validation results in Schwartz *et al.* [90]. Finally, this registration technique is now manufactured by 4-D Neuro-imaging, and during each clinical use the clinician check registration quality by superimposing the head shape points on the MRI data set. More than 300 patients have already been studied using this medical application of data fusion in Rennes University hospital.

11.7 DISCUSSION

11.7.1 Implementation of a Data Fusion Method for a Clinical Application

In the previous sections we emphasised the importance of the clinical context in data fusion methods. It is clear that the implementation of a data fusion method should be preceded by a detailed study of the clinical objectives and then its paradigm (i.e. data and clinical applications): what is the information available, and what are the clinical requirements in terms of data fusion ? Is it an application for diagnosis, planning, therapy or surgery ? Is it an inter or intra-patient application ? Are we going to model the transformation between the data with a rigid or an elastic transformation ? Are the images distorted and are we going to correct them? What is the required accuracy ? Next, the choice of a computation method depends on the type of homologous information and on the class of geometrical transformation. The choice of homologous information depends on the information available (e.g. a particular anatomical structure) in the data and on the clinical objectives. For example, is it possible to use external markers such as stereotactic frames or fiducial markers, are there any common structures, which can be easily identified such as anatomical surfaces (e.g. brain, skin, ventricles, vessels) or 3-D anatomical points, or are there no easily identifiable common structures in the data ? The choice of geometrical transformation will depend on the fusion context and determine the transformation which best explains or models the spatial variations between the different data sets. The general trend in the medical community is to prefer methods based on intrinsic homologous information, which are non-invasive. The selection of computation methods for registration (i.e. cost function and optimisation method) depends on previous choices while tools required for analysis of the registered data sets (i.e. visualisation, interaction, quantitative analysis) depend on clinical objectives.

11.7.2 Data Fusion Implementation for Clinical Use

Although many data fusion methods have been discussed in the literature for about fifteen years, there are still few industrial products or applications in clinical routine except in anatomo-functional correlation or image/patient registration (e.g. neuro-navigation, radiotherapy). There are many reasons for this, notably related to validation, additional requirements in hardware, software and human resources, and proof of clinical benefit.

As previously underlined, validation is an important step in the evaluation of a method and a prerequisite for any clinical use. It may actually be considered as a component of quality procedures. These quality procedures are not easy to define for data fusion methods: registration accuracy is difficult to quantify, there are few standardised validation data sets and no specific guidelines exist today to assist users. For example, to date there are no specific quality procedures for data fusion defined by notified bodies.

The use of multimodal data fusion in clinical routine requires a specific working environment including a computer network, image transfer and archiving capabilities (i.e. picture archiving and communication systems). Workstations and software are also required for 2-D and 3-D image processing and visualisation. Image communication standards such as DICOM [93-94], are essential to easily retrieve images of different modalities into a single workstation, on the one hand, and to retrieve information as pixel data and acquisition orientation on the other. Moreover, data fusion processes produce additional data, which have to be stored together with the original data: e.g. geometric transformation, reformatted volumes or images, segmentation results. From our experience [95] and even if many steps can be automated, data fusion procedures - from acquisition to interpretation and analysis - remain time-consuming. Moreover, data fusion is a multi-disciplinary topic embracing medicine (e.g. radiology, surgery), computer science and physics. Competence from all these fields is required during the implementation of a data fusion method. As far as clinical use is concerned, the setting up of multimodal data fusion sessions is highly desirable, to allow the different points of view of image producers and end users to be shared. Finally the real clinical benefit of data fusion has still to be demonstrated, notably through the assessment of a cost-effectiveness ratio.

11.7.3 Prospective

In this sub-section the prospects of methodology and technology, and those concerning validation and applications are discussed. Registration techniques based on a linear transformation are now quite mature, but the assumptions made to justify their use may not be fully valid. Elastic transformations allow actual phenomena to be modelled in a more realistic way, i.e. to take into account phenomena such as image distortion, patient movement, intra-operative anatomical deformations. In this chapter we have highlighted the methodological difficulties encountered in elastic matching. Elastic transformation is not morphing. This means that elastic registration should only compute geometrical and intensity variations that have to be compensated for, based on some *a priori* knowledge (i.e. explicit model of homologous features). However, this should not alter the variations corresponding to the information expected from the data fusion procedure. Improvements to the model (e.g. more realistic model of anatomical or metabolic features) and their implementations are belonging to the challenges concerning elastic matching data fusion. It is obvious that new cost functions or new optimisation methods will be defined or applied for medical image registration. One of the present trends is to combine methods (cost functions and/or optimisation methods) to improve convergence of the registration process (reducing computation time and avoiding local minima). The robustness of registration methods will also be improved by reducing sensitivity to abnormal data (e.g. pathological) thanks to, for instance, the generalisation of outlier rejection techniques. As for elastic registration, the entire data fusion procedure could be improved by the injection of *a priori* knowledge such as realistic models of anatomical or metabolic features. New technological developments will certainly produce more and more specific imaging procedures and protocols (notably in MRI), increasing the need for data fusion procedures. In addition, reduced computer costs, increased memory size and computation performance will further facilitate the deployment of computer aided data fusion procedures in clinical practice.

In this chapter we have emphasised the importance of validation to reinforce the clinical credibility of data fusion applications. It is also claimed that validation procedures should be improved, despite the difficulties raised, notably in the case of non-linear registration that remains a research issue.

Furthermore, there is a need for standardised protocols (reference data sets and methods) to provide objective validation and comparison of data fusion methods. Another important aspect concerns the definition of standardised documents to be filled in throughout validation procedures. An initial type of document would include information related to accuracy and precision values, robustness, verification of the system, computation time and complexity related characteristics. A second type of document related to daily checking procedures would record information such as residual error values and approval by clinicians for each clinical use of the system. In the same way that standards are being developed for the exchange of medical records and medical images among equipment from various manufacturers (DICOM [94]), the exchange of registration related information also requires some standardisation (e.g. registered data sets, geometrical transformations).

In the long term, data fusion tools will probably evolve towards real decision support systems including new analysis tools (e.g. quantification, statistical analysis), able to use *a priori* knowledge (e.g. knowledge databases, clinical cases) and offering communication facilities. These technological improvements will lead to the development of new clinical applications for various anatomical areas, in various medical fields (e.g. diagnosis, therapy planning, simulation, treatment, and teaching) and even to new clinical procedures. The creation of patient-specific anatomy and even physiology models will extend the capabilities of therapeutic decision-making systems. As described by Satava *et al.* [96] and Taylor *et al.* [82] the predictive medicine of the future will allow the creation of a virtual clone of the patient (i.e. a virtual patient) featuring his or her anatomical and physiological multimodal data. This virtual clone would be used for diagnosis, preoperative planning, surgical simulation, implementation and quantitative evaluation of different treatment plans and used to predict the clinical outcome of these various treatments.

11.8 CONCLUSION

The use of various complementary medical information and imaging modalities is inherent to most clinical processes and this explains why data fusion procedures were carried out even before the development of any computer-aided data fusion tools. Many methods have been designed and implemented to assist and automate data fusion processes as well as to make these processes more accurate and objective. In this chapter, it has been emphasised that the choice of appropriate data fusion methods largely depends on the clinical context, and that their validation and comparison are complex. Multimodal data fusion seems to us to have reached maturity, and wide scale deployment can be envisaged as soon as standards are available for exchanging registration information between heterogeneous platforms. Inter-patient matching techniques are still being developed but are very promising for both research purposes (e.g. in neuro-imaging) but also for clinical decision-making.

REFERENCES

1. Maintz J.A., Viergever M.A., *A review of medical image registration*, Medical Image Analysis, 1998, 2(1):1–36.
2. Maurer C.R., Fitzpatrick J.M., *A Review of Medical Image Registration*, Interactive Image-Guided Neurosurgery, Maciunas R.J. Ed., American Association of Neurological Surgeons, USA, 1993, chapter 3:17–44.
3. Van Den Elsen P.A., Pol E.J.D., Viergever M.A., *Medical image matching - a review with classification*, IEEE Eng. in Medicine and Biology Magazine, 1993, 12(1):26–39.
4. Hawkes D.J., *Algorithms for radiological image registration and their clinical application*, J. of Anatomy, 1998, 193(3):347–361.
5. Brown L.G., *A survey of image registration techniques*, ACM Computing Surveys, 1992, 24(4):325–376.
6. Duncan R., *SPECT Imaging in Focal Epilepsy*, SPECT Imaging of the Brain, Kluwer Academic Publishers, 1997, chapter 2:43-68.
7. Friston K.J., Jezzard P., Turner R., *Analysis of functional MRI time-series*, Human Brain Mapping, 1994, 2(1):153–171.
8. Worsley K.J., Friston K.J., *Analysis of fMRI time-series revisited - again*, Neuroimage, 1995, 2:173–181.
9. Lacey A.J., Thacker N.A., Jackson A., Burton E., *Locating motion artefacts in parametric fMRI analysis*, Proc. 2nd Inter. Conf. on Medical Image Computing and Computer-Assisted Interventions, 1999, Cambridge, England, Lecture Notes in Computer Science, 1679:524-532.
10. Behrenbruch C., Marias K., Armitage P., Yam M., Moore N., English R., Brady M., *MRI-mammography 2D/3D data fusion for breast pathology assessment*, Proc. 3rd Inter. Conf. on Medical Image Computing and Computer-Assisted Interventions, 2000, Pittsburgh, USA, Lecture Notes in Computer Science, Springer, 1935:307-316.
11. Boes, J.L., C.R. Meyer, *Multi-variate mutual information for registration*, Proc. 2nd Inter. Conf. on Medical Image Computing and Computer-Assisted Interventions, 1999, Cambridge, England,

Lecture Notes in Computer Science, 1679:606-612.
12. Farrell E.J., Gorniak R.J.T., Kramer E.L., Noz M.E., Maguire Jr. G.Q., Reddy D.P., *Graphical 3D medical image registration and quantification*, J. of Medical Systems, 1997, 21(3):155–172.
13. Kalfas I.H., Kormos D.W., Murphy M.A., Mc Kenzie R.L., Barnett G.H., Bell G.R., Steiner C.P., Trimble M.B., Weisenberger J.P., *Application of frameless stereotaxy to pedicle screw fixation of the spine*, Journal of Neurosurgery, 1995, 83:641–647.
14. Zana F., Klein J.C., *A multimodal registration algorithm of eye fundus images using vessels detection and Hough transform*, IEEE Trans. on Medical Imaging, 1999, 18(5):419–428.
15. Lloret D., López A., Serrat J., Villanueva J.J., *Creaseness-based computer tomography and magnetic resonance registration: comparison with the mutual information method*, J. of Electronic Imaging, 1999, 8(3):255-262.
16. Stefan H., Schneider S., Feistel H., Pawlik G., Schüler P., Abraham-Fuchs K., Schelgel T., Neubauer U., Huk W.J., *Ictal and interictal activity in partial epilepsy recorded with multichannel magnetoelectroencephalography : Correlation of Electroencephalography / Electrocorticography, magnetic resonance imaging, single photon emission computed tomography, and positron emission tomography findings*, Epilepsia, 1992, 33(5):874–887.
17. Liu A.K., Belliveau J.W., Dale A.M., *Spatiotemporal imaging of human brain activity using functional MRI constrained magnetoencephalogrphy data: Monte Carlo simulations*, Proc. Natl. Acad. Sci. USA, 1998, 95:8945–8950.
18. Maurer C.R. Jr., Hill D.L.G., Martin A.J., Liu H., McCue M., Rueckert D., Lloret D., Hall W.A., Maxwell R.E., Hawkes D.J., Truwit C.L., *Investigation of intra-operative brain deformation using a 1.5T interventional MR system: Preliminary results*, IEEE Trans. on Medical Imaging, 1998, 17(5):817–825.
19. Comeau R.M., Fenster A., Peters T.M., *Intraoperative US imaging in image-guided neurosurgery*, Radiographics, 1998, 18(4):1019–1027.
20. Olivier A., Germano I.M., Cukiert A., Peters T., *Frameless stereotaxy for surgery of the epilepsies: Preliminary experience*, J. of Neurosurgery, 1994, 81:629–633.
21. Doyle W.K., *Interactive Image-Directed Epilepsy Surgery: Rudimentary virtual reality in neurosurgery*, Interactive Technology and the New Paradigm for Healthcare, IOS Press, 1995, chapter 16:91-100.
22. Tebo S.A., Leopold D.A., Long D.M., Kennedy D.W., Zinreich S.J., *An optical 3D digitizer for frameless stereotactic surgery*, IEEE Computer Graphics and Applications, 1996, 16(1):55–64.
23. Cuchet E., Knoplioch J., Dormont D., Marsault C., *Registration in neurosurgery and neuroradiotherapy applications*, J. of Image Guided Surgery, 1995, 1:198–207.
24. Jannin P., Scarabin J.M., Rolland Y., Schwartz D., *A real 3d approach for the simulation of neurosurgical stereotactic act*, Proc. SPIE Medical Imaging VIII: Image Capture, Formatting, and Display, Newport Beach, USA, 1994, 2164:155–166.
25. Scarabin J.M., Croci S., Jannin P., Romeas R., Maurincomme E., Behague M., Carsin M., *A new concept of stereotactic room with multimodal imaging*, Proc. 13th Inter. Congress and Exhibition, Computer Assisted Radiology and Surgery, Paris, France, 1999, 841–845.
26. Simon D.A., Herbert M., Kanade T., *Techniques for fast and accurate intrasurgical registration*, J. of Image Guided Surgery, 1995, 1(1):17–29.
27. Weese J., Penney G.P., Desmedt P., Buzug T.M., Hill D.L.G., Hawkes D.J.H., *Voxel-based 2-D/3-D registration of fluoroscopy images and CT scans for image-guided surgery*, IEEE Trans. on Information Technology in Biomedicine, 1997, 1(4):284–293.
28. Weese J., Buzug T. M., Lorenz C., Fassnacht C., *An approach to 2D/3D registration of a vertebra in 2D X-ray fluoroscopies with 3D CT images*, Lecture Notes in Computer Science, 1997, 1205:119-128.
29. Betting F., Feldmar J., *3D-2D projective registration of anatomical surfaces with their projections*, Proc. 14th Information Processing in Medical Imaging Conf., 1995, Ile de Berder, France, 3:275-286.
30. Evans A.C., Collins D.L., Mills S. R., Brown E. D., Kelly R.L., *3D statistical neuroanatomical models from 305 MRI volumes*, Proc. IEEE Nuclear Science Symp. and Medical Imaging Conf., San Francisco, USA, 1993, 1813–1817.
31. Collins D.L., Holmes C.J., Peters T.M., Evans A.C., *Automatic 3D model-based neuroanatomical*

segmentation, Human Brain Mapping, 1995, 3(3):190–208.
32. Ashburner J., Friston K.J., *Multimodal image coregistration and partitioning - a unified framework*, NeuroImage, 1997, 6(3):209–217.
33. Mazziotta J.C., Toga A.W., Evans A.C., Fox P., Lancaster J.L., *A probabilistic atlas of human brain: Theory and rationales for its development*, Neuroimage, 1995, 2:89–101.
34. Friston K.J., Ashburner J., Poline J.B., Frith C.D., Heather J.D., Frackowiak R.S.J., *Spatial registration and normalization of images*, Human Brain Mapping, 1995, 2:165–189.
35. Guimond A., Roche A., Ayache N., Meunier J., *Multimodal brain warping using the demons algorithm and adaptive intensity corrections*, INRIA, France, Report N°RR-396, 1999.
36. Barillot C., *Fusion de données et imagerie 3D en médecine*, Thèse d'habilitation à diriger des recherches, IRISA, Université de Rennes, France, 1999.
37. Talairach J., Tournoux P., *Co-Planar stereotactic atlas of the human brain*, Georg Thieme Verlag, 1988.
38. Collins D.L., Le Goualher G., Evans A.C., *Non-linear cerebral registration with sulcal constraints*, Proc. 1st Inter. Conf. on Medical Image Computing and Computer-Assisted Interventions, 1998, Cambridge, USA, Lecture Notes in Computer Science, 1496:974–984.
39. Van Den Elsen P., Maintz A. Pol E., Viergever M., *Automatic registration of CT and MR brain images using correlation of geometrical features*, IEEE Trans. on Medical Imaging, 1995, 14(2):384-396.
40. Subsol G., Thirion J.P., Ayache N., *Steps towards automatic building of anatomical atlases*, SPIE Proc. 3rd Conf. on Visualization in Biomedical Computing, 1994, Rochester, USA, 2359:435-446.
41. Faugeras O., *Three-Dimensional Computer Vision: A Geometric Viewpoint*, MIT Press, 1993.
42. Bookstein F.L., *Principal warps: Thin-plate splines and the decomposition of deformations*, IEEE Trans. on Pattern Analysis and Machine Intelligence, 1989, 11(6):567–585.
43. Bookstein F.L., *Thin-plate splines and the atlas problem for biomedical images*, Proc. 12th Conf. on Information Processing in Medical Imaging, 1991, Wye, England, Lecture Notes in Computer Science, 511:326–342.
44. Sandor S., Leahy R., *Surface-based labelling of cortical anatomy using a deformable atlas*, IEEE Trans. on Medical Imaging, 1997, 16(1):41–54.
45. Thompson P., Toga A., *A surface-based technique for warping three-dimensional images of the brain*, IEEE Trans. on Medical Imaging, 1996, 15(4):402–417.
46. Szeliski R., Lavallée S., *Matching 3-D anatomical surfaces with non-rigid deformations using octree splines*, Inter. J. of Computer Vision, 1996, 18(2):176–186.
47. Collins D.L., Peters T.M., Evans A.C., *An automated 3D non-linear deformation procedure for determination of gross morphometric variability in human brain*, SPIE Proc. 3rd Conf. on Visualization in Biomedical Computing, 1994, Rochester, USA, 2359:180–190.
48. Hellier P., Barillot C., Mémin E., Pérez P., *Medical image registration with robust multigrid techniques*, Proc. 2nd Inter. Conf. on Medical Image Computing and Computer-Assisted Interventions, 1999, Cambridge, England, Lecture notes in Computer Science, 1679:680–687.
49. Thirion J.P., *Non-rigid matching using demons*, Proc. Conf. on Computer Vision and Pattern Recognition, Los Alamitos, USA, IEEE Computer Society Press, 1996, 245-251.
50. Thirion J.P., *Image matching as a diffusion process: An analogy with Maxwell's demons*, Medical Image Analysis, 1998, 2(3):243–260.
51. Bajcsy R., Kovacic S., *Multiresolution elastic matching*, Computer Vision, Graphics, and Image Processing, 1989, 46(1):1–21.
52. Gee J., Reivicj M., Bajcsy R., *Elastically deforming 3D atlas to match anatomical brain images*, J. of Computed Assisted Tomography, 1993, 17(2):225–236.
53. Christensen G.E., Rabbitt R.D., Miller M.I., Joshi S.C., Grenander U., Coogan T.A., Van Essen D.C., *Topological properties of smooth anatomic maps*, Proc. 14th Inter. Conf. on Information Processing in Medical Imaging, 1995, Dordrecht, Netherlands, 3:101–112.
54. Miller M.I., Christensen G.E., Amit Y.A., Grenander U., *Mathematical textbook of deformable neuroanatomies*, Medical Sciences, 1993, 90:11944–11948.
55. Borgefors G., *A new distance transformation approximating the Euclidean distance*, Proc. 8th Inter. Conf. on Pattern Recognition, 1986, Paris, France, 336–338.
56. Borgefors G., *Hierarchical chamfer matching: A parametric edge matching algorithm*, IEEE

Trans. on Pattern Analysis and Machine Intelligence, 1988, 10(6):849–865.

57. Pelizzari C.A., Chan G. T.Y., Spelbring D.R., Weichselbaum E. E., Chen C.T., *Accurate three-dimensional registration of CT, PET and/or MR images of the brain*, J. of Computer Assisted Tomography, 1989, 13(1):20–26.
58. Roche A., Malandain G., Ayache N., Pennec X., *Multimodal image registration by maximization of the correlation ratio*, INRIA, France, Report N°RR-3378, 1998.
59. Roche A., Malandain G., Pennec X., Ayache N., *The correlation ratio as a new similarity measure for multimodal image registration*, Proc. 1st Inter. Conf. on Medical Image Computing and Computer-Assisted Intervention, 1998, Cambridge, USA, Lecture Notes in Computer Science, 1496:1115-1124.
60. Woods R.P., Mazziotta J.C., Cherry S.R., *MRI-PET registration with automated algorithm*, J. of Computer Assisted Tomography, 1993, 17(4):536–546.
61. Maes F., Collignon A., Vandermeulen D., Marchal G., Suetens P., *Multimodality image registration by maximization of mutual information*, IEEE Trans. on Medical Imaging, 1997, 16(2):187–198.
62. Wells III W.M., Viola P., Atsumi H., Nakajima S., Kikinis R., *Multi-modal volume registration by maximization of mutual information*, Medical Image Analysis, 1996, 1(1):35–51.
63. Roche A., Malandain G., Ayache N., *Unifying maximum likelihood approaches in medical image registration*, INRIA, France, Report N°RR-3741, 1999.
64. De Castro E., Morandi C., *Registration of translated and rotated images using finite Fourier transforms*, IEEE Trans. Pattern Analysis and Machine Intelligence, 1987, 9:700-703.
65. Pluim J.P.W., Maintz J.B.A., Viergever M.A., *Interpolation artefacts in mutual information-based image registration*, Computer Vision and Image Understanding, 2000, 77(2):211–232.
66. Maes F., Vendermeulen D., Suetens P., *Comparative evaluation of multiresolution optimization strategies for multimodality image registration by maximisation of mutual information*, Medical Image Analysis, 1999, 3(4):373–386.
67. Black M.J., Rangarajan A., *On the unification of line processes, outlier rejection, and robust statistics with applications in early vision*, Inter. J. of Computer Vision, 1996, 19(1):57–91.
68. Nikou C., Heitz F., Armspach J.P., Namer I.J., Grucker D., *Registration of MR/MR and MR/SPECT brain images by fast stochastic optimization of robust voxel similarity measures*, Neuroimage, 1998, 8:30–43.
69. Press W.H., Teukolsky S.A., Vetterling W.T., Flannery B.P., *Numerical Recipes in C*, Cambridge University Press, 1992.
70. Jannin P., Bouliou A., Scarabin J.M., Barillot C., Lubert J., *Visual matching between real and virtual images in image guided neurosurgery*, SPIE Proc. on Medical Imaging: Image Display, 1997, 3031:518–526.
71. Jannin P., Grova C., Schwartz D., Barillot C., Gibaud B., *Visual qualitative comparison between functional neuro-imaging (MEG, fMRI, SPECT)*, Proc. 13th Computer Assisted Radiology and Surgery Conf., 1999, Paris, France, 238-243.
72. Viergever M.A., Maintz J.B.A., Stokking R., *Integration of functional and anatomical brain images*, Biophysical Chemistry, 1997, 68(1-3):207–219.
73. Socoloinsky D.A., Wolff L.B., *Image fusion for enhanced visualization of brain imaging*, SPIE Proc. on Medical Imaging: Image Display, 1999, 3658:352– 362.
74. Noordmans H.J., Van der Voort H.T.M., Rutten G.J.M., Viergever M.A., *Physically realistic visualization of embedded volume structures for medical image data*, SPIE Proc. on Medical Imaging: Image Display, 1999, 3658:613– 620.
75. Colin A., Boire J.-Y., *MRI-SPECT fusion for the synthesis of high resolution 3d functional brain images : A preliminary study*, Computer Methods and Programs in Biomedicine, 1999, 60(2):107–116.
76. Hawkes D.J., Hill D.L.G., Lehmann E.D., Robinson G.P., Maisey M.N., Colchester A.C.F., *Preliminary work on the interpretation of SPECT images with the aid of registered MR images and an MR derived 3D neuro-anatomical atlas*, 3D Imaging in Medicine, Springer Verlag, 1990, 241-251.
77. O'Brien T.J., O'Connor M.K., Mullan B.P., Brinkmann B.H., Hanson D., Jack C.R., So E.L., *Subtraction ictal SPECT co-registered to MRI in partial epilepsy: Description and technical*

validation of the method with phantom and patient studies, Nuclear Medicine Communications, 1998, 19:31–45.

78. Lundervold A., Storvik G., *Segmentation of brain parenchyma and cerebrospinal fluid in multispectral magnetic resonance images*, IEEE Trans. on Medical Imaging, 1995, 14(2):339-349.
79. Brinkmann B.H., O'Brien T.J., Aharon S., O'Connor M.K., Mullan B.P., Hanson D.P., Robb R.A., *Quantitative and clinical analysis of SPECT image registration for epilepsy studies*, The J. of Nuclear Medecine, 1999, 40(7):1098–1105.
80. Le Goualher G., Procyk E., Collins D.L., Venugopal R., Barillot C., Evans A.C., *Automated extraction and variability analysis of sulcal neuroanatomy*, IEEE Trans. on Medical Imaging, 1999, 18(3):206–217.
81. Friston K.J., Holmes A.P., Worsley K.J., Poline J.B., Frith C.D., Frackowiak R.S.J., *Statistical parametric maps in functional imaging: A general linear approach*, Human Brain Mapping, 1995, 2:189–210.
82. Taylor C.A., Draney M.T., Ku J.P., Parker D., Steele B.N., Wang K., Zarins C.K., *Predictive medicine: Computational techniques in therapeutic decision-making*, Computer aided surgery, 1999, 4:231–247.
83. Sumanaweera T.S., Adler J.R., Napel S., Glover G.H., *Characterization of spatial distortion in magnetic resonance imaging and its implications for stereotactic surgery*, Neurosurgery, 1994, 35(4):696–704.
84. Collins D.L., Zijdenbos A.P., Kollokian V., Sled J., Kabani N.J., Holmes C.J., Evans A.C., *Design and construction of a realistic digital brain phantom*, IEEE Trans. on Medical Imaging, 1998, 17(3):463–468.
85. West J., Fitzpatrick J.M., Wang M.Y., Dawant B.M., Maurer C.R., Kessler R.M., Maciunas R.J., Barillot C., Lemoine D., Collignon A., Maes F., Suetens P., Vandermeulen D., Van Den Elsen P.A., Napel S., Sumanaweera T.S., Harkness B., Hemler P.F., Hill D.L.G., Hawkes D.J., Studholme C., Maintz J.B.A., Viergever M.A., Malandin G., Pennec X., Noz M.E., Maguire G.Q., Pollack M., Pellizzari C.A., Robb R.A., Hanson D., Woods R., *Comparison and evaluation of retrospective intermodality image registration techniques*, SPIE Proc. on Medical Imaging: Image Processing, 1996, 2710:332–347.
86. West J., Fitzpatrick J.M., Wang M.Y., Dawant B.M., Maurer C.R., Kessler R.M., Maciunas R.J., Barillot C., Lemoine D., Collignon A., Maes F., Suetens P., Vandermeulen D., Van Den Elsen P.A., Napel S., Sumanaweera T.S., Harkness B., Hemler P.F., Hill D.L.G., Hawkes D.J., Studholme C., Maintz J.B.A., Viergever M.A., Malandin G., Pennec X., Noz M.E., Maguire G.Q., Pollack M., Pellizzari C.A., Robb R.A., Hanson D., Woods R., *Comparison and evaluation of retrospective intermodality brain image registration techniques*, J. of Computer Assisted Tomography, 1997, 21(4):554–566, 1997.
87. Van Bemmel J.H., Musen M.A., *Handbook of Medical Informatics*, Springer Verlag, 1997.
88. Vieth J., Kober H., Weise E., Daun A., Moeger A., Friedrich, Pongratz H., *Functional 3d localization of cerebrovascular accidents by magnetoencephalography (MEG)*, Neurological Research, 1992, 14:132–134.
89. Schwartz D., *Localisation Des Générateurs Intra-Cérébraux de L'activité MEG et EEG: Evaluation de la Précision Spatiale et Temporelle*, PhD thesis, Université de Rennes 1, France, 1997.
90. Schwartz D., Poiseau E., Lemoine D., Barillot C., *Registration of MEG/EEG data with MRI : Methodology and precision issues*, Brain Topography, 1996, 9(2):101–116.
91. Schwartz D., Badier J.M., Bihouee P., Bouliou A., *Evaluation of a new MEG-EEG spatio-temporal approach using realistic sources*, Brain Topography, 1999, 11(4):279–289.
92. Barillot C., Schwartz D.P., Le Goualher G., Gibaud B., Scarabin J.M., *Representation of MEG/EEG data in a 3D morphological environment*, Computer Assisted Radiology, Proc. of the Inter. Symp. on Computer and Communication Systems for Image Guided Diagnosis and Therapy, 1996, Paris, France, 249–254.
93. Yoo T.S., Ackerman M.J., Vannier M., *Toward a common validation methodology for segmentation and registration algorithms*, Proc. 3rd Inter. Conf. on Medical Image Computing and Computer-Assisted Interventions, 2000, Pittsburgh, USA, Lecture Notes in Computer Science, Springer, 1935:422-431.

94. DICOM Committee, *Digital Imaging and Communications in Medecine (DICOM)*, ACR-NEMA Standard PS3.1-9, 1993.
95. Jannin J., Fleig O.J., Seigneuret E., Grova C., Morandi X., Scarabin J.M., *A data fusion environment for multimodal and multi-informational neuro-navigation*, J. of Computer Aided Surgery, 2000, 5(1):1-10.
96. Satava R.M., Jones S.B., *Current and future applications of virtual reality for medicine*, Proc. of IEEE, 1998, 86(3):484–489.

APPENDIX NDT DATA FUSION ON THE WEB

Below is a list of Internet Web sites that may be useful to readers wishing to know more about NDT data fusion and information fusion.

Homepage of Huadong
Correlation of accelerometer and microphone data in the coin tap test
http://www.cs.cmu.edu/~whd/

Integrated Predictive Diagnostics
Development of data fusion systems for condition monitoring
Research at Pennsylvania State University, USA
http://www.arl.psu.edu/areas/soa/conditionmaint.html

International Society of Information Fusion
General information, conferences, terminology
http://www.inforfusion.org/

Homepage of Xavier E. Gros
Research in NDT data fusion
http://www.geocities.com/xgros/

Homepage of Pierre Jannin
Medical applications of NDT data fusion, case studies and course on multimodal data fusion
http://sim3.univ-rennes1.fr/fusion/clinicalcase.htm
http://sim3.univ-rennes1.fr/users/jannin/COURSMMgb/index.htm

Bundesanstalt für Materialforschung und -prüfung (BAM)
Homepage of the TRAPPIST project for transfer, processing and interpretation of 3-D NDT data in a standard environment
http://trappist.kb.bam.de/

University of Drexel
Distributed detection and decision fusion
http://dflwww.ece.drexel.edu/~bai/research/ciss97/
http://dflwww.ece.drexel.edu/research/

University of Tokyo
Description of sensor fusion projects
http://www.k2.t.u-tokyo.ac.jp/~sfoc/projects/SensorFusion/

Cooperative Research Centre for Signal Processing and Information Processing
Inference and decision analysis, sensor and information fusion
http://www.cssip.edu.au/

NDTnet
Extended abstract about NDE data fusion in the aerospace industry
http://www.ndt.net/abstract/asntf97/053.htm

Georgia Institute of Technology, Intelligent Control Systems Laboratory
Projects on fuzzy logic, neural networks, integration, optimization, and control of industrial processes, diagnostic/sensors and sensing strategies, etc.
http://icsl.marc.gatech.edu/

Georgia Institute of Technology, Intelligent Control Systems Laboratory
Signal processing and decision making techniques (neural networks, fuzzy logic), feature extraction and selection for characterisation, detection and prediction for medical applications
http://130.207.32.221/Projects/eeg/brochure_eeg.htm

University of Iowa
Geometrically accurate fusion of image data from biplane angiography and intravascular ultrasound
http.//www.icaen.uiowa.edu/~ceig/RESEARCH/BSC-research.html
Image guided neurosurgery
http://procrustes.radiology.uiowa.edu/~guest/neurosurgery/sld001.htm

Helsinki University of Technology
Data fusion and neural networks in complex models
http://zeus.hut.fi/research/Impress/

VTT Technical Research Centre of Finland
A European funded research project to develop, validate and evaluate biosignal interpretation methods applied in neuro-monitoring
http://www.vtt.fi/tte/samba/projects/ibis/overview.html

Homepage of Robert W. Cox
Processing, analysing, and displaying functional MRI (FMRI) data to map human brain activity
http://varda.biophysics.mcw.edu/~cox/index.html

CREATIS at INSA Lyon
Fusion between models and ultrasonic data for medical applications
http://www.insa-lyon.fr/Insa/Laboratoires/creatis.html

Aptronix Inc.
Fuzzy logic modelling
http://www.aptronix.com/

University of Michigan, Department of Radiology
Digital image processing
http://www.med.umich.edu/dipl/research.html

Vanderbilt University, Computer Science Image Processing Laboratory
The Retrospective Registration Evaluation Project
http://cswww.vuse.vanderbilt.edu/~image/registration

National Library of Medicine
The Visible Human Project
http://www.nlm.nih.gov/research/visible/visible_human.html

University College London, The Wellcome Department of Cognitive Neurology
Statistical Parametric Mapping
http//www.fil.ion.ucl.ac.uk/spm

UCLA, Laboratory of Neuro Imaging
Automated image registration
http://bishopw.loni.ucla.edu/AIR3/

Homepage of Seth Hutchinson
Artificial intelligence, robotics, computer vision
http://www.beckman.uiuc.edu/faculty/hutchins.html

US National Geophysical Data Center
What is data fusion ? Satellite image fusion.
http://web.ngdc.noaa.gov/seg/tools/gis/fusion.shtml

Homepage of Robert Pawlak
Data and sensor fusion resources
http://users.aol.com/mlkienholz/df.html

NASA Langley Research Center
The autonomous NDE agent will make use of data fusion technology
http://nesb.larc.nasa.gov/NESB/ndetasks/2000/autoagent.html

Italian Aerospace Research Centre
NDE data fusion research project
http://www.cira.it/research/nde/nde.htm

National Research Council Canada
Structures, materials and propulsion laboratory
http://www.nrc.ca/iar/smp_nde-e.html

The URL addresses listed above were correct at the time of printing.

INDEX

MIX
Papier aus verantwortungsvollen Quellen
Paper from responsible sources
FSC® C105338

If you have any concerns about our products,
you can contact us on
ProductSafety@springernature.com

In case Publisher is established outside the EU,
the EU authorized representative is:
Springer Nature Customer Service Center GmbH
Europaplatz 3, 69115 Heidelberg, Germany

Printed by Libri Plureos GmbH
in Hamburg, Germany